NOV 2012

D1594795

DATE DUE

JAN 2 9 2013			

DEMCO 38-296

Which Numbers are Real?

© *2012 by the Mathematical Association of America, Inc.*

Library of Congress Catalog Card Number 2012937493

Print edition ISBN 978-0-88385-777-9

Electronic edition ISBN 978-1-61444-107-6

Printed in the United States of America

Current Printing (last digit):
10 9 8 7 6 5 4 3 2 1

Which Numbers are Real?

Michael Henle
Oberlin College

Published and Distributed by
The Mathematical Association of America

Council on Publications and Communications
Frank Farris, *Chair*

Committee on Books
Gerald M. Bryce, *Chair*

Classroom Resource Materials Editorial Board
Gerald M. Bryce, *Editor*

Michael Bardzell
Jennifer Bergner
Diane L. Herrmann
Philip P. Mummert
Barbara E. Reynolds
Susan G. Staples
Philip D. Straffin
Cynthia J Woodburn
Holly S. Zullo

CLASSROOM RESOURCE MATERIALS

Classroom Resource Materials is intended to provide supplementary classroom material for students—laboratory exercises, projects, historical information, textbooks with unusual approaches for presenting mathematical ideas, career information, etc.

101 Careers in Mathematics, 2nd edition edited by Andrew Sterrett

Archimedes: What Did He Do Besides Cry Eureka?, Sherman Stein

Calculus: An Active Approach with Projects, Stephen Hilbert, Diane Driscoll Schwartz, Stan Seltzer, John Maceli, and Eric Robinson

The Calculus Collection: A Resource for AP and Beyond, edited by Caren L. Diefenderfer and Roger B. Nelsen

Calculus Mysteries and Thrillers, R. Grant Woods

Conjecture and Proof, Miklós Laczkovich

Counterexamples in Calculus, Sergiy Klymchuk

Creative Mathematics, H. S. Wall

Environmental Mathematics in the Classroom, edited by B. A. Fusaro and P. C. Kenschaft

Excursions in Classical Analysis: Pathways to Advanced Problem Solving and Undergraduate Research, by Hongwei Chen

Exploratory Examples for Real Analysis, Joanne E. Snow and Kirk E. Weller

Geometry From Africa: Mathematical and Educational Explorations, Paulus Gerdes

Historical Modules for the Teaching and Learning of Mathematics (CD), edited by Victor Katz and Karen Dee Michalowicz

Identification Numbers and Check Digit Schemes, Joseph Kirtland

Interdisciplinary Lively Application Projects, edited by Chris Arney

Inverse Problems: Activities for Undergraduates, Charles W. Groetsch

Laboratory Experiences in Group Theory, Ellen Maycock Parker

Learn from the Masters, Frank Swetz, John Fauvel, Otto Bekken, Bengt Johansson, and Victor Katz

Math Made Visual: Creating Images for Understanding Mathematics, Claudi Alsina and Roger B. Nelsen

Mathematics Galore!: The First Five Years of the St. Marks Institute of Mathematics, James Tanton

Ordinary Differential Equations: A Brief Eclectic Tour, David A. Sánchez

Oval Track and Other Permutation Puzzles, John O. Kiltinen

A Primer of Abstract Mathematics, Robert B. Ash

Proofs Without Words, Roger B. Nelsen

Proofs Without Words II, Roger B. Nelsen

Rediscovering Mathematics: You Do the Math, Shai Simonson

She Does Math!, edited by Marla Parker

Solve This: Math Activities for Students and Clubs, James S. Tanton

Student Manual for Mathematics for Business Decisions Part 1: Probability and Simulation, David Williamson, Marilou Mendel, Julie Tarr, and Deborah Yoklic

Student Manual for Mathematics for Business Decisions Part 2: Calculus and Optimization, David Williamson, Marilou Mendel, Julie Tarr, and Deborah Yoklic

Teaching Statistics Using Baseball, Jim Albert

Visual Group Theory, Nathan C. Carter

Which Numbers are Real?, Michael Henle

Writing Projects for Mathematics Courses: Crushed Clowns, Cars, and Coffee to Go, Annalisa Crannell, Gavin LaRose, Thomas Ratliff, Elyn Rykken

MAA Service Center
P.O. Box 91112
Washington, DC 20090-1112
1-800-331-1MAA   FAX: 1-301-206-9789

Introduction

Alternative reality

The real numbers are fundamental. Although mostly taken for granted, they are what make possible all of mathematics from high school algebra and Euclidean geometry through the calculus and beyond, and also serve as the basis for measurement in science, industry, and ordinary life. In this book we study alternative systems of numbers: systems that generalize and extend the reals yet stay close to the fundamental properties that make the reals central to so much mathematics.

By an alternative number system we mean a set of objects that can be combined using two operations, addition and multiplication, and that share some significant algebraic and geometric properties with the real numbers. Exactly what these properties are is made clear in Chapter One. We are not concerned with numeration, however. A numeration system is a means of giving names to numbers, for example, the decimal system for writing real numbers. We go beyond numeration to describe number systems that include numbers different from ordinary numbers including multi-dimensional numbers, infinitely small and infinitely large numbers, and numbers that represent positions in games.

Although we present some eccentric and relatively unexplored parts of mathematics, each system that we study has a well-developed theory. Each system has applications to other areas of mathematics and science, in particular to physics, the theory of games, multi-dimensional geometry, formal logic, and the philosophy of mathematics. Most of these number systems are active areas of current mathematical research and several were discovered relatively recently. As a group, they illuminate the central, unifying role of the reals in mathematics.

Design of this book

This book is designed to encourage readers to participate in the mathematical development themselves. The proofs of many results are either contained in problems or depend on results proved in problems. The problems should be read at least, if not worked out.

With two exceptions the chapters are independent and can be read in any order. The exceptions are that the first two chapters contain essential background for the rest of the book and that Chapter Four depends somewhat on the proceeding chapter.

Use of this book

This book presents material that, in addition to being of general interest to mathematics students, is appropriate for an upper level course for undergraduates that can serve as introduction to or sequel to a course in advanced calculus. Alternatively, it can take the place of a course in the foundations of the real number system, or be given as an upper level seminar emphasizing different methods of proof. Prerequisites are standard sophomore level courses: discrete mathematics, multivariable calculus, and linear algebra. A course in advanced calculus or foundations of analysis would also be useful.

The goal

The goal is to present some interesting, even exotic, mathematics. I hope to convey a sense of the immense freedom available in mathematics, where even in a mundane and well-established area such as the real numbers, alternatives are always possible.

Acknowledgements

This book owes a tremendous debt to previous expositors in this area to whose work this book is closely tied, in particular, to the work of Elwyn Berlekamp, Errett Bishop, John H. Conway, Richard Guy, James Henle, and Eugene Kleinberg.

Contents

Introduction		**vii**
I	**THE REALS**	**1**
1	**Axioms for the Reals**	**3**
	1.1 How to Build a Number System	3
	1.2 The Field Axioms	12
	1.3 The Order Axioms	18
	1.4 The Completeness Axiom	24
2	**Construction of the Reals**	**35**
	2.1 Cantor's Construction	36
	2.2 Dedekind's Construction of the Reals	43
	2.3 Uniqueness of the Reals	46
	2.4 The Differential Calculus	50
	2.5 A Final Word about the Reals	52
II	**MULTI-DIMENSIONAL NUMBERS**	**55**
3	**The Complex Numbers**	**57**
	3.1 Two-Dimensional Algebra and Geometry	57
	3.2 The Polar Form of a Complex Number	62
	3.3 Uniqueness of the Complex Numbers	66
	3.4 Complex Calculus	70
	3.5 A Final Word about the Complexes	76

4 The Quaternions — 77
- 4.1 Four-Dimensional Algebra and Geometry — 77
- 4.2 The Polar Form of a Quaternion — 82
- 4.3 Complex Quaternions and the Quaternion Calculus — 86
- 4.4 A Final Word about the Quaternions — 94

III ALTERNATIVE LINES — 95

5 The Constructive Reals — 97
- 5.1 Constructivist Criticism of Classical Mathematics — 97
- 5.2 The Constructivization of Mathematics — 103
- 5.3 The Definition of the Constructive Reals — 109
- 5.4 The Geometry of the Constructive Reals — 114
- 5.5 Completeness of the Constructive Reals — 118
- 5.6 The Constructive Calculus — 119
- 5.7 A Final Word about the Constructive Reals — 124

6 The Hyperreals — 125
- 6.1 Formal Languages — 126
- 6.2 A Language for the Hyperreals — 134
- 6.3 Construction of the Hyperreals — 138
- 6.4 The Transfer Principle — 142
- 6.5 The Nature of the Hyperreal Line — 150
- 6.6 The Hyperreal Calculus — 156
- 6.7 Construction of an Ultrafilter — 162
- 6.8 A Final Word about the Hyperreals — 170

7 The Surreals — 171
- 7.1 Combinatorial Games — 172
- 7.2 The Preferential Ordering of Games — 178
- 7.3 The Arithmetic of Games — 184
- 7.4 The Surreal Numbers — 188
- 7.5 The Nature of the Surreal Line — 192
- 7.6 More Surreal Numbers — 196
- 7.7 Analyzing Games with Numbers — 200
- 7.8 A Final Word about the Surreals — 204

Bibliography — 205

Index — 209

About the Author — 219

Part I

THE REALS

What makes a number system a number system? In this book the real numbers serve as the standard with which other number systems are compared. To be called a number system a mathematical system must share most if not all of the fundamental properties of the reals.

What are the fundamental properties of the reals? We use a set of properties (or laws or axioms) that characterize the reals completely, meaning that any mathematical system with these properties is the same as the reals. Such a set of properties for a particular mathematical object is called a categorical axiom system. Many such systems are known. A famous one for plane geometry goes back to Euclid, although a correct and complete categorical axiom system for Euclidean geometry was formulated only late in the 19th century. In this later period and on into the 20th century there has been tremendous interest in axiom systems and their application to all areas of mathematics.

Part One of this book describes a categorical axiom system for the reals. Chapter One lists the axioms of this system. Chapter Two constructs the reals (from the rational numbers), and shows that they satisfy the axioms presented in Chapter One. In addition, we prove that any mathematical system satisfying these axioms is identical (more technically, isomorphic) to the reals.

A categorical axiom system is a powerful tool. The one we describe is used in this book to analyze and compare number systems. Given any system we ask: which axioms for the reals does it satisfy? The answer reveals how close the new system is to the standard set by the reals themselves. Of course, any proposed system that is genuinely different from the reals cannot satisfy all the axioms, for then it would be the reals.

In principle, we could survey systematically all possible number systems, first by finding those satisfying all but one of the axioms of the reals, then finding systems satisfying all but two axioms, and so forth. We cannot accomplish this ambitious program; only partial results are known! Even these encompass a veritable ocean of important modern mathematics. However, we can present the most important mathematical systems that satisfy all but a few of the axioms.

Important Note: We assume that the reader is already familiar with two kinds of numbers: the integers (i.e., the whole numbers, positive, negative and zero), and the rational numbers (the common fractions). The basic properties of these numbers are assumed. The theory of the reals and other number systems will be based on them.

1

Axioms for the Reals

1.1 How to Build a Number System
Equivalence relations

In this section we describe a process that is used to construct many number systems and other mathematical systems as well. It is used to construct about half of the number systems in this book. This process uses the concept of an equivalence relation. Here is the definition:

Definition. Let S be a set and $\sim$ a relationship that may or may not hold between two elements of S. A relation on S satisfying these properties:

    (a) For every a in S, $a \sim a$,   — **reflexivity**

    (b) For a, b in S, if $a \sim b$, then $b \sim a$,   — **symmetry**

    (c) For a, b, c in S, if $a \sim b$, and $b \sim c$, then $a \sim c$.   — **transitivity**

is called an **equivalence relation** on S.

The three properties—reflexivity, symmetry, and transitivity—are the axioms of equivalence relations, sometimes called laws of equivalence. An example of an equivalence relation is the relation of equality. Every set S has this relation. The idea of an equivalence relation is an abstraction of equality, and, as we will soon see, every equivalence relation can be turned into the relationship of equality on some set.

We see many examples of equivalence relations connected with number systems later. Here are a few diverse examples.

Problems

1. Let S be the set of all triangles in the plane. For A, B in S, set $A \equiv B$ if A and B are congruent. Explain why $\equiv$ is an equivalence relation for S.

    (*Hint*: We are asked to show that "$\equiv$" has the three properties: reflexivity, symmetry and transitivity. For example, is $\equiv$ reflexive? According to the definition, this means: Is every triangle congruent to itself? Congruence of triangles means that corresponding sides are equal (in some order) and corresponding angles. Therefore, any triangle is congruent to itself: the sides are equal to themselves, and the angles are equal to themselves. The answer is yes: congruence is reflexive.)

2. Consider two further relations between triangles:

    $A \approx B$ if A and B are similar triangles,

    $A \angle B$ if at least one angle of A equals an angle of B.

    Are $\approx$ and $\angle$ equivalence relations?

3. Let S be the set of people in Australia. For A, B in S define $A \approx B$ if A and B have the same birthday. Is $\approx$ an equivalence relation?

4. Let a set U be given. Let S be the set of subsets of U. For A, B in S, define $A \Leftrightarrow B$ if $|A| = |B|$ (i.e., A and B have the same number of elements). Is $\Leftrightarrow$ an equivalence relation?

5. Let $\mathbb{Z}$ be the set of integers, positive, negative and zero. Let p be a positive integer. For x and y in $\mathbb{Z}$ define $x \equiv y \pmod{p}$ if p divides $x - y$ with remainder zero. Is $\equiv$ an equivalence relation?

6. Let $\mathbb{F}$ be the set of all symbols of the form a/b where a and b are integers and b is not zero. (For the purpose of this exercise forget that / is sometimes used for fractions.) Define a relation $\sim$ for these symbols $a/b \sim c/d$ if $ad = bc$. Is $\sim$ an equivalence relation?

7. Find a set S and a relation on S that is
    (a) reflexive, but not symmetric or transitive,
    (b) symmetric, but not reflexive or transitive,
    (c) transitive, but not reflexive or symmetric,
    (d) reflexive and symmetric but not transitive,
    (e) symmetric and transitive but not reflexive,

1.1. How to Build a Number System

(f) transitive and reflexive but not symmetric.

(*Hint*: Good examples of relationships can be found outside mathematics. For example, "cousinhood" has an interesting combination of properties. It is symmetric, but not reflexive (i.e., one is not one's own cousin).

Examples of relations can be found on any finite set. Furthermore, if S is finite, a relation on S can be diagrammed as in Figure 1.1.1. The set S there has only three elements, A, B, and C. The arrows indicate the relationships that exist among the elements of S. If the relation is symbolized by "$\neg$", say, then $A \neg B$, $B \neg C$ and $C \neg C$. It is not reflexive (A is not related to itself), nor symmetric ($A \neg B$, but not $B \neg A$), nor transitive ($A \neg B$ and $B \neg C$, but not $A \neg C$).

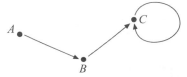

Figure 1.1.1. A simple relation on a small set.

Equivalence classes

To use equivalence relations to construct number systems, we need another definition.

Definition. Let S be a set with an equivalence relation $\sim$. For each a in S let
$$S_a = \{b | b \sim a\}.$$
The set S_a is called the **equivalence class** containing a. The set of all the equivalence classes is written $S/\sim$.

Usually, an equivalence relation is defined on a set S precisely in order to study the set $S/\sim$ of equivalence classes. $S/\sim$ is the new set built by introducing the equivalence relation. In this book this new set will usually be a number system but the same construction is used for many other mathematical concepts.

Proof is one of the most difficult of mathematical concepts (see the shaded text below). The next exercises provide the opportunity to supply proofs of some fundamental facts about equivalence classes.

> **What is a proof?**
>
> Originally, to prove something meant to try it out, or to test it (leaving open, of course, the possibility of failure). For example, a baker *proofs* the yeast before baking, and Christopher Marlowe's passionate shepherd says, "Come live with me and be my love, and we will all the pleasures prove."
>
> In mathematics, on the other hand, a proof is supposed to be an argument that leaves no possible doubt of its truth. Mathematical proofs are not supposed to fail. Some do, however, because of human error, because standards of proof evolve so that what was once certain becomes doubtful or even wrong, and because of the intrinsic subtlety of the whole process.
>
> No one has ever set down exactly what makes a mathematical proof a proof. No one knows infallibly what a proof is. On the other hand, attacking proofs is easy: every statement that does not carry utter conviction is vulnerable to criticism.
>
> Perhaps proof, like Zen Buddhism (as described by Alan Watts) "can have no positive definition. It has to be suggested by saying what it is not, somewhat like a sculptor reveals an image by the act of removing pieces of stone from a block." Unfortunately, this makes finding proofs something like "a game in which the rules [have] been partially concealed."
>
> How do we cope with this situation? When a proof is proposed, it is read by other mathematicians. As more and more people study it, test it, and work with it, it gradually achieves acceptance.
>
> In summary, although we can't say exactly what a proof is, it's always possible to *proof* proofs, that is, to test them. To test your proof, let others read it. See whether they are convinced.

Problems

8. Describe a typical equivalence class for the equivalence relations described in problems 1–5.

   (*Example*: For congruence for triangles (Problem 1), a typical equivalence class consists of all triangles of a particular shape and size.)

1.1. How to Build a Number System

9. Prove that all the elements of S_a are equivalent to each other.
   (*Hint*: Let b and c be elements of S_a. Then $b \sim a$ and $c \sim a$. Use the laws of equivalence to draw the desired conclusion.)
10. Prove that if $a \sim b$, then $S_a = S_b$.
11. Prove that every element a of S is a member of exactly one equivalence class.

Building number systems with equivalence classes: a discussion

Let's start with one of the simplest number systems: the integers, $\mathbb{Z}$, consisting of the whole numbers—positive, negative, and zero—i.e., $\mathbb{Z} = \{0, \pm 1, \pm 2, \ldots\}$. Besides numbers, $\mathbb{Z}$ comes with the two operations of addition and multiplication, so we can add and multiply integers—always obtaining, as it happens, another integer as a result.

Addition is an invertible operation in $\mathbb{Z}$; that is, given integers p and q, it is possible to un-add p to q. For example, un-adding 5 to 7 gives 2. (Check this: take 2, add back the 5, and note that we do get 7.) "Un-adding," of course, is usually called "subtraction". Un-adding 5 is adding -5 to 7, the number -5 being the **additive inverse** of 5. The point here is that un-adding is not a problem in the integers: -5 is an integer. It is inside $\mathbb{Z}$ with all the other integers. Every integer has an additive inverse in $\mathbb{Z}$.

Multiplication is not invertible. We cannot un-multiply 7 by 5 because there is no integer x such that multiplying it by 5 gives 7. In symbols, the equation $5x = 7$ has no integer solution. We can fix this by building a new number system, the rational numbers $\mathbb{Q}$, where un-multiplication (except by zero!) is possible. (For the sake of this discussion, please forget, for a page or two, that you know about the rational numbers already.)

Towards the goal of building from the integers a number system in which un-multiplication is possible, let $\mathbb{F}$ be the set of all symbols of the form p/q where p and q are integers and q is not zero. Call the symbols p/q **fractions**. Think of p/q as representing the result of un-multiplying p by q (if only this were possible), but at the moment p/q is a pure symbol symbolizing nothing.

These so-called fractions name the numbers in our new system, the rational numbers. But the set of fractions is not the same as the set of rational numbers because many fractions name the same number.

How do I know that many different fractions name the same number?

Here is one argument that suggests this must be true. In $\mathbb{Z}$, multiplication is commutative. In our new number system, let us agree that un-multiplication will be commutative, a natural property we might desire any operation possess. We stipulate that commutativity hold for fractions and hope that nothing bad happens, where 'bad' means a contradiction appears.

It happens that un-multiplication is sometimes possible even within $\mathbb{Z}$. If an integer p factors, say $p = bc$, then we can un-multiply p by b, getting c. For example, we can un-multiply 12 by 4 getting 3—all within the integers. Ah, but suppose that p factors in two different ways: $p = ad = bc$. In this situation, if we un-multiply p by b we get c, and if we then un-multiply c by d, we get the fraction c/d. If un-multiplication is commutative, this must be the same as un-multiplying p by d first (getting a), and then un-multiplying a by b getting a/b. Therefore, if un-multiplication is commutative, and $ad = bc$, then the fractions a/b and c/d name the same rational number. Let us write $a/b \sim c/d$ in this situation. This is an equivalence relation for $\mathbb{F}$ (see problem 6), and the new number system we seek is the set of equivalence classes $\mathbb{Q} = \mathbb{F}/\sim$.

In summary, 3/6, 4/8 and 23/46, for example, are all names for the same rational number, namely 1/2 or one-half (to use its simplest fractional name). We are so used to this particular equivalence relation that we call the fractions 4/8 and 23/46 equal without thinking anything of it. They are not equal, however, at least not as elements of $\mathbb{F}$. (Look at them: 4/8 and 23/46 are different!) They are equal in the set $\mathbb{Q}$, though, since they belong to the same equivalence class.

The view of equivalence relations and equivalence classes that emerges from this discussion is this: An equivalence relation on S supplies S with a new definition of equality, since equivalence relations have the same algebraic properties as equality. If we use the equivalence relation in place of equality, then the original set S functions as a set of names for a new set of objects: the set of equivalence classes: $S/\sim$.

An essential feature of our equivalence relation for fractions is that a/b and c/d can be tested for equivalence by a calculation (does $ad = bc$?) carried out entirely within $\mathbb{Z}$ without using any un-multiplications. Un-multiplication is still a fictitious operation since we have not completed construction of the new number system $\mathbb{Q}$.

What else is needed to complete the creation of $\mathbb{Q}$? Quite a lot of work! First is the problem of definition: the operations of addition and multiplication must be defined for rational numbers, and it must be proved that they satisfy the same basic properties in $\mathbb{Q}$ as in $\mathbb{Z}$, for example the commutative

1.1. How to Build a Number System

and associative laws. Then there is the embedding problem: verifying that the old system $\mathbb{Z}$ is contained inside the new system $\mathbb{Q}$. This means that inside $\mathbb{Q}$ a copy of the integers $\mathbb{Z}$ must be found that behaves, so far as addition and multiplication are concerned, exactly as the original integers do. Finally, it must be shown that un-multiplication is possible in $\mathbb{Q}$ since this was the reason for the construction of $\mathbb{Q}$ in the first place.

We will not tackle all this right now as we will have similar tasks to complete while constructing the real numbers and various alternative real number systems.

(You are now allowed to remember that you already know about the rational numbers.)

The problem of definition

One unfinished element of the construction of the rational numbers is worth pursuing here: the problem of how to define operations on a set of equivalence classes.

For the symbols a/b, addition is defined by

$$a/b + c/d = (ad + bc)/(bd).$$

This is the well-known law for the addition of fractions based on a common denominator bd of the two fractions. It defines addition not for the rational numbers themselves, however, but only for their symbolic names from $\mathbb{F}$. Can this symbolic addition be applied to the equivalence classes? It can, if it is well-defined, meaning that if

$$a/b \sim a'/b',$$

and

$$c/d \sim c'/d',$$

then

$$a/b + c/d \sim a'/b' + c'/d'.$$

In other words, for an operation defined on a set of symbols to be *well-defined* on the equivalence classes, it must be proved that equivalent symbols combined with equivalent symbols give equivalent results. The issue of well-definedness arises whenever a mathematical system is defined using equivalence classes.

In praise of names

It may seem that we belittle the fractions a/b by saying that they only are names for the rational numbers. If so, we should correct this impression right away. It is true that, for example, $20/25$ and $80/100$ and $580/725$, and so on are names for an underlying ideal rational quantity $(4/5)$, but for practical purposes we have to have these names in order to do anything with the rationals. In general, when a new number system $S/\sim$ is constructed using an equivalence relation, S is not discarded. We need those names.

Problems

12. Prove that the symbolic addition of fractions is well-defined for the equivalence classes $\mathbb{Q} = \mathbb{F}/\sim$.

13. For the symbolic fractions a/b of the set $\mathbb{F}$, define multiplication by setting
$$(a/b)(c/d) = (ac)/(bd).$$
Prove that multiplication is well-defined.

14. The fractions in $\mathbb{F}$ of the form $p/1$ serve as embedded copies of the integers in the rationals.
    (a) Verify that equivalence on $\mathbb{F}$ for fractions of this form is the same as equality.
    (b) Verify that addition of fractions of this form is the same as the usual addition of integers.
    (c) Verify that multiplication of fractions of this form is the same as the usual multiplication of integers.

Where do number systems really come from?

The reader at this point may object that the process by which we have obtained the rational numbers from the integers cannot possibly be how the rational numbers were actually discovered. A few words about this now will place the theoretical description of number systems, to which this book is devoted, in the context of their historical development.

The discovery and development of the rational numbers was a long process beginning before the dawn of written history. Because fractions are essential for commercial and astronomical calculation, they were discovered

1.1. How to Build a Number System

by many civilizations, including the ancient Egyptians and Babylonians. Over the centuries, rules for calculation with specific fractions slowly gave way to more general procedures. Many notations were invented and used. The current, standard notation was invented several times, most recently by Hindu mathematicians sometime before 600 AD. They used it without the bar separating numerator from denominator, which was added by Arab mathematicians later.

In the history of the discovery of the rational numbers and other number systems, theory follows calculation. Thus, a theory of number was established, by the Greeks, only after centuries of systematic calculation. Euclid's *Elements* (c. 300 BC) contains the first formal exposition of a theory of numbers (as well as a theory of geometry). One of the most important Greek discoveries about rational numbers, incidentally, is that there are irrational numbers. This is attributed to the Pythagorean school (c. 400 BC). After the Greeks, the theory of number systems is dominated by efforts to come to terms with various troublesome numbers: the negative, the irrational, and the complex—among others. In all this, the positive rationals remain uncontroversial.

Although much was learned in the post-Grecian era about calculation with numbers, not much progress was made toward understanding their theoretical nature until the nineteenth century. Then rapid progress was made as part of a general program undertaken by European mathematicians to place the ideas of the calculus on a firm foundation. All the classical number systems received definitive treatment in the nineteenth century: the complexes first (c. 1800: Gauss and others), then the reals (1852–83: Dedekind and Cauchy), then the rationals (c. 1854–67: described by Bolzano and Hankel as a system of numbers closed under addition, subtraction, multiplication, and division), and, finally, the integers (1889: Peano). The equivalence relation $\sim$ on $\mathbb{F}$ appears first in an 1895 algebra text by Weber. The order in which the theory of these number systems developed is the reverse of the order in which they were discovered for computation and the reverse of the order in which they are usually studied.

The theoretical development of number systems made possible the invention of many more systems in the 19th and the 20th centuries. In this book we describe five of the most important of these. They appear in historical order: the complex numbers (1800 and earlier), the quaternions (1843), the constructive reals (1870–1930), the (1961), and the surreals (1970). Each of these systems was created in response to a computational challenge: the complexes in order to solve quadratic and higher degree equations, the quaternions in order to calculate with vectors and compute transformations of three-dimensional space, the constructive reals in order to deal with perceived limits on the nature of computation, the hyperreals in order to calculate with infinitesimals, and the surreals in order to evaluate positions in combinatorial games. Their discovery is a legacy of theoretical developments of the 19th century including the description of the rationals as $\mathbb{F}/\sim$. For more on the history of number systems, see references [A1] and [A2]. For more on the history of mathematics in general, see [A3].

Summary

Number systems have a long practical history but their theory developed only rather recently. One important theoretical development is the invention of equivalence relations, which permit the elements of a set to serve as names for a new set constructed from the old. Equivalence relations appear in many parts of mathematics. In this book they are used to create new number systems.

1.2 The Field Axioms

The algebra of the reals

In this and the next two sections we describe a categorical axiom system for the reals. The axioms of this system divide conveniently into three subsets: axioms for the algebra of the reals, axioms for the geometry of the reals, and a final axiom called completeness.

Algebra is the most familiar of these; it is the natural place to start.

Definition. Let S be a set with two operations called **addition** and **multiplication** and written with the usual signs. We assume that the set S is **closed** under these operations, so that applying the operations to two elements of S produces another element of S.

Then S is a **field** if addition and multiplication have the properties:

(a) For a, b, c in S,

$$(a+b)+c = a+(b+c), \quad \text{— associativity of addition}$$

and

$$(ab)c = a(bc). \quad \text{— associativity of multiplication}$$

(b) For a, b in S,

1.2. The Field Axioms

$$a + b = b + a, \quad \text{— commutativity of addition}$$

and

$$ab = ba. \quad \text{— commutativity of multiplication}$$

(c) For a, b in S,

$$a(b + c) = ab + ac. \quad \text{— distributivity}$$

(d) There are distinct, special elements 0 (**zero**) and 1 (**one**) in S such that for any b in S,

$$0 + b = b + 0 = b, \quad \text{— additive identity}$$

and

$$1b = b1 = b. \quad \text{— multiplicative identity}$$

(e) For any b in S, there is an element $-b$ in S so that

$$b + (-b) = 0, \quad \text{— additive inverse}$$

and for any b in S, except 0, there is an element c in S so that

$$bc = 1, \quad \text{— multiplicative inverse}$$

Properties (a)–(e) are the **field axioms**. They summarize the algebra of the real numbers, the rational numbers, and many other number systems. They were first collected together under the term 'field' by Weber in 1893. See reference [C7].

Assumption. Let $\mathbb{Q}$ be the set of all rational numbers, that is numbers of the form p/q, where p and q are integers and q is not zero. We assume that $\mathbb{Q}$ is a field.

Problems

1. Let $\mathbb{Z}$ be the set of all integers. $\mathbb{Z}$ is not a field. List the axiom(s) not satisfied by $\mathbb{Z}$.

2. Let n be a positive integer, and consider the set $\mathbb{Z}/n$ of integers modulo n, that is $\mathbb{Z}/n$ is the set of equivalence classes of $\mathbb{Z}$ under the equivalence relation $x \equiv y \pmod{n}$ if n divides $x - y$ with remainder zero. How many equivalence classes are there in $\mathbb{Z}/n$?

3. Show that addition and multiplication are well-defined on $\mathbb{Z}/n$, that is if $a \equiv b \pmod{n}$ and $c \equiv d \pmod{n}$, then $a + c \equiv b + d \pmod{n}$ and $ac \equiv bd \pmod{n}$.

4. Write out the complete addition and multiplication tables for $\mathbb{Z}/2$, $\mathbb{Z}/3$, $\mathbb{Z}/4$, $\mathbb{Z}/5$, and $\mathbb{Z}/6$.

5. For what values of n is $\mathbb{Z}/n$ a field?

   (*Hint*: Problem 14 below describes the integral domain property, a property that all fields have. For which n does $\mathbb{Z}/n$ have the integral domain property.)

6. For rational numbers a and b define $a \sim b$, if $(a - b)$ is an integer. Is $\sim$ an equivalence relation? If so, are addition and multiplication well-defined? If so, is $\mathbb{Q}/\sim$ a field?

7. Here is another relation on the rationals. Let a, b, c, d, be integers, with b and d not zero, and define $a/b \leftrightarrow c/d$ if $ad - bc$ is even. Is $\leftrightarrow$ an equivalence relation? If so, are addition and multiplication well-defined? If so, is $\mathbb{Q}/\leftrightarrow$ a field?

8. Let $\mathbb{Q}(\bullet)$ be the set of symbols of the form $a + b\bullet$, where a and b are rationals. (Call $\bullet$ 'blob', then $\mathbb{Q}(\bullet)$ is 'cue-blob'.) Define addition and multiplication in $\mathbb{Q}(\bullet)$ by setting

   $$(a + b\bullet) + (c + d\bullet) = (a + c) + (b + d)\bullet,$$

   and

   $$(a + b\bullet)(c + d\bullet) = (ac + 3bd) + (bc + ad)\bullet.$$

   Prove that $\mathbb{Q}(\bullet)$ is a field.

   (*Hint*: To find the multiplicative inverse of the element $(a + b\bullet)$ in $\mathbb{Q}(\bullet)$ solve the equation

   $$(a + b\bullet)(c + d\bullet) = 1 + 0\bullet,$$

   for c and d by solving a pair of linear equations.)

9. Let $\mathbb{Q}(x)$ be the set of all rational functions

   $$f(x) = \frac{p(x)}{q(x)},$$

   where $p(x)$ and $q(x)$ are polynomials with rational coefficients and $q(x)$ is not the zero polynomial. Let addition and multiplication in $\mathbb{Q}(x)$ be the usual addition and multiplication of functions. Show that $\mathbb{Q}(x)$ is a field.

1.2. The Field Axioms

Solving equations with the field axioms

The field axioms justify many algebraic techniques for solving equations. The problems show how.

Problems

10. Prove the **cancellation law of addition**: in any field S, if $a + b = a + c$, then $b = c$.
    (*Hint*: Use an inverse operation.)
11. Prove the **cancellation law of multiplication**: in any field S, if $ab = ac$ and $a \neq 0$, then $b = c$.
12. Explain how to solve the linear equation $ax + b = c$ for x, where a, b and c are elements in a field S and $a \neq 0$, using the cancellation laws of addition and multiplication.
13. Solve these linear equations:
    (a) $\frac{1}{2}x + \frac{1}{3} = \frac{4}{9}$, for x in the field $\mathbb{Q}$,
    (b) $2 + 3 \bullet a + b + 1 - 4 \bullet = 3 + \bullet$, for $a + b \bullet$ in the field $\mathbb{Q} \bullet$,
    (c) $\left(\dfrac{x^2 - 1}{x}\right) f(x) + \left(\dfrac{x^3 - 2x + 3}{x^2 + 1}\right) = \left(\dfrac{x^4 + 4x^2 - 5}{x^2 + 1}\right)$, for $f(x)$ in the field $\mathbb{Q}(x)$.
14. Prove the **integral domain property**: in a field S, if $ab = 0$, then either $a = 0$ or $b = 0$.
15. Explain how to solve the quadratic equation $x^2 + (a + b)x + ab = 0$, where a and b are elements in a field S and x is unknown, using the integral domain property.
16. Solve these quadratic equations:
    (a) $x^2 + 4x - 60 = 0$, for x in the field $\mathbb{Q}$,
    (b) $(a + b \bullet)^2 + (2 + 2 \bullet)(a + b \bullet) + (3 + 2 \bullet) = 0$, for $(a + b \bullet)$ in the field $\mathbb{Q}(\bullet)$,
    (c) $f(x)^2 - \dfrac{x^3 - 4x^2 + x - 2}{x^2 - 4} f(x) - \dfrac{x^3 - x^2 + x}{x^2 - 4} = 0$, for $f(x)$ in the field $\mathbb{Q}(x)$.

    (*Hint*: Factor the left-hand side of these equations in the usual way by factoring the constant terms, e.g., 60 in equation (a). Remember that $3 = \bullet^2$, in $\mathbb{Q}(\bullet)$.)

> **How to find a proof**
>
> There is no sure way to find proofs, but there are general principles, called **heuristics**, useful in all problem-solving situations. They include:
>
> 1. *Talk to yourself.* Ask: What do I know already about this? Ask: Is there extra information I need and where can I get it? Ask: What if I could prove such-and-such, then can I prove what I want?
> 2. *Work out a plan.* Divide the proof into stages or cases that can be tackled separately.
> 3. *Be flexible.* Work both forward and backward, i.e., from the beginning or from the end. It may even be possible to start in the middle and work toward both ends.
> 4. *Draw a picture.* Also doodle, invent your own symbols, make up notation, build models.
> 5. *Successively refine.* Don't expect your first thoughts to be perfect in all detail. Scribble down ideas for later polishing. Go over the proof several times with a critical eye asking yourself: Is it convincing?
> 6. *Indirect proof.* If necessary, try proof by contradiction.

The next exercises describe results that hold in all fields. Try heuristics on the proofs.

Problems

17. Let n be a positive integer, and let a be an element of a field. Define na by setting

    $$na = a + a + a + \cdots + a. \quad \text{— exactly } n \text{ } a\text{'s}$$

    Prove that

    $$n(a+b) = na + nb,$$
    $$(n+m)a = na + ma,$$

1.2. The Field Axioms

and
$$(na)b = a(nb) = n(ab),$$
where n and m are positive integers, and a and b are field elements.

18. Prove that the additive inverse of an element of a field is unique.

19. Prove that the multiplicative inverse of a non-zero element of a field is unique.

20. Prove that $-a = (-1)a$ and $-(-a) = a$ where a is an element of a field.

    (*Hint*: Use the uniqueness of additive inverses.)

21. For a nonzero element b in a field S, let b^{-1} stand for the multiplicative inverse of b. Prove that
$$(-b)^{-1} = -(b^{-1}),$$
$$(b^{-1})^{-1} = b,$$
and
$$(ab)^{-1} = b^{-1}a^{-1}.$$

    (*Hint*: Use the uniqueness of multiplicative inverses.)

When are two fields the same?

The answer is provided by the concept of isomorphism.

Definition. Two fields S and T are **(field) isomorphic** if there is a one-to-one, onto function $i : S \to T$ (i.e., a bijection) such that
$$i(a + b) = i(a) + i(b),$$
and
$$i(ab) = i(a)i(b).$$
The function i is called a **(field) isomorphism**.

Informally, the isomorphism of two fields, S and T, means that every algebraic operation in one of them (for example, $a + b$ in S) is imitated by a parallel operation in the other (that is, $i(a) + i(b)$ in T).

Problems

22. Show that the identity function $i: S \to S$ is an isomorphism.
23. Show that the fields $\mathbb{Z}/p$ for different p are not isomorphic to each other.
24. Prove that field isomorphism is an equivalence relation on a set of fields.

Summary

The field axioms set forth algebraic laws that, as we shall see, are satisfied by the real numbers. We therefore expect them to be satisfied by any number system that is an alternative to the reals. The field axioms are the basis for many of the usual algebraic techniques of solving equations. The concept of isomorphism makes precise when two fields have identical algebraic structure.

1.3 The Order Axioms

The geometry of the reals

Geometrically the reals are a straight line. This is expressed by order axioms.

Definition. A field S is **(linearly) ordered** if there is a subset S^+ of S satisfying:

(a) If a and b are in S^+, then so are $a + b$ and ab,

(b) If a is in S, then exactly one of the following is true: $a = 0$, a is in S^+, or $-a$ is in S^+ (**the trichotomy law**).

The elements of S^+ are called **positive**, while if $-x$ is in S^+ we call x **negative**.

Two ordered fields S and T are **(order) isomorphic** if there is a field isomorphism $i: S \to T$ such that a is in S^+ if and only if $i(a)$ is in T^+.

To say that the reals form a line means that given two real numbers, x and y, we can tell which is further along the line, that is, whether $x < y$ or $y < x$. Thus, the linear nature of the reals takes the algebraic form of working with inequalities. Here is how inequalities are defined in an ordered field.

1.3. The Order Axioms

Definition. Let S be any ordered field. Let a and b be elements of S. We write
$$a < b,$$
when $b - a$ is positive. We write $a \leq b$ when $b - a$ is positive or zero.

Assumption. The rationals $\mathbb{Q}$ are an ordered field.

The order axioms are not satisfied in many fields. Most of the fields we have introduced are not ordered.

Problems

1. Prove that 1 is positive in an ordered field.

2. Show that the fields $\mathbb{Z}/p$ are not ordered.

   (*Hint*: 1 is positive by the previous exercise, so $1 + 1$ is also positive. Argue that 0 is positive.)

3. Let $(a + b\bullet)$ be an element in the field $\mathbb{Q}(\bullet)$. Let us say that $(a + b\bullet)$ is a positive element of $\mathbb{Q}(\bullet)$ if either a and b are both positive, or $a \leq 0$, $b > 0$ and $a^2 < 3b^2$, or $a > 0$, $b \leq 0$ and $a^2 > 3b^2$. Show that $\mathbb{Q}(\bullet)$ is an ordered field.

4. For an element $f(x) = p(x)/q(x)$ in the field $\mathbb{Q}(x)$ of all rational functions the degree of f is the degree of p minus the degree of q. For example the degree of
$$f(x) = \frac{x^2 - 1}{x^3 + x + 1}$$
is $2 - 3 = -1$. Show that defining the positive elements of $\mathbb{Q}(x)$ to be those whose degree is positive does not make $\mathbb{Q}(x)$ an ordered field.

Solving inequalities with the order axioms

The order axioms justify the usual algebraic laws of inequalities, including those in the next problems.

Problems

5. Prove the trichotomy law for inequalities: given a and b in an ordered field S, exactly one of $a = b$, $a < b$, and $b < a$ is true.

6. Prove the transitive law of inequalities: for a, b, and c in an ordered field S, if $a < b$ and $b < c$, then $a < c$.
7. Prove the law of addition of inequalities: in an ordered field if $a < b$ and $c < d$, then $a + c < b + d$.
8. Prove in any ordered field that if $a > b$ then $-b > -a$.
9. In an ordered field, prove that the product of two negatives is positive, while the product of a negative and a positive is negative.
10. Let S be an ordered field. Prove that $1 + 1 \neq 0$. Deduce that between any two elements of S there are an infinite number of other elements.

    (*Hint*: Define 2 by $1 + 1$. Given two field elements, a and b, show that their average $(a + b)/2$ lies between them.)

Absolute value and distance

Every ordered field has absolute values. They are used to define distance. Here is the definition.

Definition. Let S be an ordered field and let a and b be elements of S. The **absolute value** is defined by

$$|a| = \begin{cases} a & \text{if } a \text{ is positive or zero,} \\ -a & \text{otherwise.} \end{cases}$$

The **distance** between a and b is defined as $|a - b|$.

The absolute value $|a - b|$ is the distance between a and b measured in units from the ordered field itself. The next few problems develop some properties of absolute values and distance.

Problems

11. Prove that in any ordered field the absolute value function satisfies:

    $$|a + b| \leq |a| + |b|, \quad \text{— \textbf{triangle inequality}}$$

    and

    $$|na| = n|a|,$$

    where n is a positive integer.

1.3. The Order Axioms

12. Prove that in any ordered field:
    (a) The distance of any point to itself is the zero of the field.
    (b) The distance from a to b equals the distance from b to a.
    (c) Given three points, a, b, and c, the distance from a to c is less than or equal to the sum of the distance from a to b plus the distance from b to c.

13. In an ordered field, show that if $|a| < k$, then $-k < a < k$. Prove that if $|b - a| < k$, then $a - k < b < a + k$.

Embedding Theorems

The next theorem is the first of many embedding theorems we prove. It describes how the natural numbers are embedded inside every ordered field.

Theorem 1.3.1. *Let S be an ordered field. For any natural number n define*

$$i(n) = 1 + 1 + \cdots + 1. \qquad \text{— exactly } n \text{ ones}$$

where 1 is the multiplicative identity of S. Then i satisfies:
(a) $i(n + m) = i(n) + i(m)$,
(b) $i(nm) = i(n)i(m)$,
(c) i is one-to-one.

Embedding the natural numbers in an ordered field S is executed by a function. The theorem says that every integer n has a copy of itself, $i(n)$, inside S. Parts (a) and (b) express the fact that given two natural numbers (n and m) their copies (that is, $i(n)$ and $i(m)$) in S behave arithmetically exactly like the originals. Part (c) says that the embedding is strict, i.e., one-to-one.

As this is our first formal theorem, we give a complete proof, the heart of which is an argument by mathematical induction.

Proof. The definition of $i(n)$ implies that $i(n + 1) = i(n) + 1$. This makes proof by induction work.

Proof of (a). (By mathematical induction) Let $s(n)$ be the statement "$i(n + m) = i(n) + i(m)$ for all m."

Base Case: The statement $s(1)$ is "$i(m + 1) = i(m) + 1$". This is clear from the definition of $i(n)$.

> **Mathematical induction**
>
> This powerful technique is used to prove a sequence of propositions, one for each natural number. Let $s(n)$ be some statement that depends on the value of the natural number n. Proof by induction consists of two steps:
>
> *Base case*: Prove $s(1)$.
>
> *Inductive case*: Assume that $s(n)$ has been proved. Use this assumption to prove $s(n+1)$.
>
> What justifies this kind of proof? From a modern point-of-view, the validity of mathematical induction is a fundamental assumption concerning the nature of the natural numbers. Still, what do we to say to someone who challenges a proof by induction? It is not very satisfying simply to assert that we assume that this kind of proof is valid.
>
> If someone challenges proof by mathematical induction, ask them in return, "If you question my proof (that $s(n)$ is true for all n), then for what number n do you think $s(n)$ is false?" The rest of the conversation might proceed along the following lines.
>
> "You say you doubt $s(7)$? But $s(1)$ is true, right, because I proved the base case. And you know that $s(2)$ follows from $s(1)$ because I proved the inductive case. Similarly $s(3)$ follows from $s(2)$, and $s(4)$ from $s(3)$, and so forth, and so on. Eventually, we reach 7. So $s(7)$ is true."
>
> Eventually, we reach any particular natural number. This is one of Peano's axioms for the natural numbers and is what ultimately justifies mathematical induction.

Inductive Case: Assume that "$i(n+m) = i(n) + i(m)$ for all m" is true for a specific n. Then for $n+1$ we have

$$\begin{aligned}
i\big((n+1)+m\big) &= i(n+m+1) \\
&= i(n+m) + 1 &&\text{— by definition of } i(n+m) \\
&= i(n) + i(m) + 1 &&\text{— by assumption} \\
&= i(n+1) + i(m). &&\text{— by definition of } i(n)
\end{aligned}$$

Thus we have proved the statement $s(n+1)$. This completes the proof of (a).

1.3. The Order Axioms

Proof of (b). (By mathematical induction) Let $s(n)$ be the statement "$i(nm) = i(n)i(m)$ for all m."

Base Case: The statement $s(1)$ is "$i(m) = i(m)$" which is clear.

Inductive Case: Assume that "$i(nm) = i(n)i(m)$ for all m" is true for a specific n. Then for $n+1$ we have

$$
\begin{aligned}
i((n+1)m) &= i(nm + m) & &\text{— distributive law for integers} \\
&= i(nm) + i(m) & &\text{— by (a)} \\
&= i(n)i(m) + i(m) & &\text{— by assumption} \\
&= (i(n) + 1)i(m) & &\text{— distributive law in } S \\
&= i(n+1)i(m). & &\text{— definition of } i(n)
\end{aligned}
$$

Thus we have proved the statement $s(n+1)$. This completes the proof of (b).

Proof of (c). Suppose that $i(n) = i(m)$. This means that

$$1 + 1 + \cdots + 1 \ (n \text{ times}) = 1 + 1 + \cdots + 1. \ (m \text{ times}) \quad (*)$$

As long as there is a 1 on each side of this equation we can cancel it from both sides. Eventually (after many cancellations, perhaps), one side or the other of the equation will be reduced to zero. If this happens on just one side, we get

$$1 + 1 + \cdots + 1 \ (q \text{ times}) = 0.$$

Because 1 is positive, the sum $1 + 1 + \cdots + 1$ (q times) is also positive. By the trichotomy law for S, it is impossible that a positive number be zero. Therefore, as we cancel 1's from both sides of $(*)$, both sides will become zero simultaneously, proving that the two sides of $(*)$ have the same number of 1's or that $m = n$. This proves that i is one-to-one. □

This embedding of the natural numbers in an ordered field can be extended to the integers as described in the next theorem.

Theorem 1.3.2. *Let S be an ordered field. Let i be the embedding defined in Theorem 1.3.1. Extend i to negative integers and 0 by setting $i(0) = 0$ and $i(-n) = -i(n)$. Then i still has the properties*

(a) $i(n + m) = i(n) + i(m)$,

(b) $i(nm) = i(n)i(m)$, and

(c) i is one-to-one.

Problems

14. Where in the proof of the Theorem 1.3.1 did we use the fact that 1 is the multiplicative identity of S?

15. Prove Theorem 1.3.2.

    (*Hint*: The three parts of this theorem have already been proved for positive numbers. To prove them for negative numbers, survey all possible cases. The result of Problem 13 in §1.2 may be useful. To prove (c) show that if $i(n) = i(m)$, then $i(n-m) = 0$.)

Summary

The order axioms are the properties that characterize a straight line. They are the basis for the usual algebraic techniques for solving inequalities. Order isomorphism makes precise when two ordered fields have identical ordered structure.

An ordered field has an embedded copy of the integers. These copies of the integers combine according to the same laws of addition and multiplication as the integers themselves. From now on we always assume that an ordered field contains the integers and use n (rather than $i(n)$) to stand for the copy of n in that field.

1.4 The Completeness Axiom

What makes the reals unique?

The rationals $\mathbb{Q}$ and the reals $\mathbb{R}$ are ordered. So are many fields in between them. For example, the set of all numbers of the form $a + b\sqrt{3}$, where a and b are rational numbers, is an ordered field that contains the rationals but does not contain all reals. This is the field $\mathbb{Q}(\bullet)$ of problem 8 in §1.2 and problem 3 of §1.3. What makes the real numbers unique among ordered fields is another property (or law or axiom) called completeness, which amounts to the assertion that the field has no gaps or holes. The purpose of this section is to make this precise and intelligible.

There is no rational number in $\mathbb{Q}$ whose square is 2. This is what we mean by a hole: the rationals are missing the square root of 2, a number we can approximate to as many places as we like, that occupies, it seems,

1.4. The Completeness Axiom

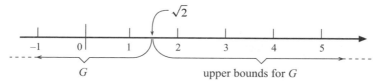

Figure 1.4.1. A set bounded above and some of its upper bounds.

a definite place along a number line, but cannot be expressed as a fraction p/q where p and q are integers.

Consider the set of rational numbers less than $\sqrt{2}$. To be specific, let

$$G = \{a \mid a \text{ is rational and } a^2 < 2\}.$$

This set contains rationals whose square is arbitrarily close to 2, but no number whose square equals 2. (See Figure 1.4.1.)

The set G is bounded above in the sense of the following definition. We use the bounds of G to get at the missing $\sqrt{2}$.

Definition. Let S be an ordered field. A subset G of S is **bounded above** if there is an element k of S such that $a < k$ for all a in G. Then k is called an **upper bound** for G.

An element b of S is a **least upper bound** (**lub** for short) for a set G if b is an upper bound for G and $b \leq k$ for all upper bounds k of G.

For example, G in Figure 1.4.1 is bounded above by $k = 3$ (see problem 2). Although G is bounded above, G does not have a least upper bound in the rationals. In the reals, however, G has a least upper bound, which, after we construct the reals, turns out to be $\sqrt{2}$).

Problems

1. Prove that there is no rational number p/q whose square is 2.

   (*Hint*: Assume that p/q is in lowest terms, that is, that p and q have no common factor. Use the fact that 2 is prime.)

2. Prove that the set $G = \{a \mid a \text{ is rational and } a^2 < 2\}$ has a rational upper bound.

   (*Hint*: 100, for example, is an upper bound. Try a contrapositive proof.)

3. Let H be a subset of an ordered field S, and suppose that H has least upper bound b. For a positive integer n prove that

(a) $b + 1/n$ is not in H, but is an upper bound for H,
(b) $b - 1/n$ is not an upper bound for H,
(c) there is an element g of H such that $b - 1/n < g \le b$.

4. Verify that G (Figure 1.4.1) has no rational least upper bound without using the square root of 2, i.e., give a proof using only rational numbers.

5. Find an example of a set of rationals that has a rational least upper bound.

6. Define **greatest lower bound (glb)** and find an example of a set of rationals that is bounded below but has no rational greatest lower bound.

7. Find and describe a set of rationals whose least upper bound is $-\sqrt{5}$. Do this without explicitly using the number $-\sqrt{5}$.

8. Find and describe a set of rationals whose least upper bound is e. Do this without explicitly using the number e.
 (*Hint*: Use an infinite series for e.)

9. Find and describe a set of rationals whose least upper bound is π. Do this without explicitly using the number π.

Completeness

The completeness axioms uses the terminology of boundedness and least upper bounds.

Completeness Axiom. *An ordered field S is called (**order**) **complete** if every non-empty subset of S with an upper bound has a least upper bound in S.*

If we accept that the least upper bound of G is $\sqrt{2}$, then by insisting that the reals be complete we force $\sqrt{2}$ to be a real number.

Problems

10. Show that if a field S is complete, then a subset of S that is has a greatest lower bound.

11. In a complete, ordered field S, for each a in S^+ there is a unique integer n such that $n \le a < n + 1$ called the **greatest integer in** a and denoted $\lceil a \rceil$. Prove the existence and uniqueness of $\lceil a \rceil$.

1.4. The Completeness Axiom

Order completeness and the Archimedean property

To understand completeness, we introduce some related properties, the first of which is the Archimedean property. Suppose we have obtained a ruler marked in units of length a in order to measure a line of length b, where a and b are positive elements from an ordered field. (See Figure 1.4.2.) The Archimedean property asserts that we *can* measure the length b using a ruler marked in units of a provided the ruler is long enough. More precisely, the Archimedean property says that there is a natural number n such that if the ruler is at least n units of length a long, then it can measure any length up to b. Here is the formal statement.

Definition. An ordered field S is **Archimedean** if, given two positive numbers a and b, there is a positive integer n such that $b < na$.

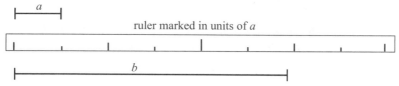

Figure 1.4.2. Units, a ruler, and a length to be measured by it.

It may seem obvious that a ruler marked in one length can measure any other length but this cannot be proved using the axioms of an ordered field alone. A further axiom or assumption is needed.

The Archimedean property is linked to the existence (or not) of infinite elements. As it happens, the axioms for an ordered field don't prevent it from containing infinitely large elements. (An **infinitely large** element is an element b of an ordered field S larger than all the natural numbers $n1$ embedded in S.) The hyperreals and the surreals, for example (see Chapters 6 and 7), contain infinitely large numbers. In an Archimedean field, however, this cannot happen. There, for any positive b, there is a positive integer n such that $b < n1$.

The Archimedean property is also linked with completeness: order completeness implies that a field is Archimedean:

Theorem 1.4.1. *A complete field is Archimedean.*

Problems

12. In an Archimedean field S, prove that for a in S^+ there is an integer n such that $0 < 1/n < a$. This says that an Archimedean field does not contain elements that are infinitely small.

13. Prove Theorem 1.4.1.

    (*Detailed hint*: Let $T = \{n \in S \mid n \text{ is a positive integer and } na = b\}$. Show that T is bounded above by $1 + b/a$. If T is empty there is nothing to prove (why?). Otherwise, T is non-empty and bounded above, so that by the completeness of S, T has a least upper bound, say M. Now $M - 1$ is not an upper bound for T (why?), so there exists an integer k in T such that
    $$M - 1 < k \leq M.$$
    The integer $n = k + 1$ is greater than M, hence not in T (why?). Therefore $na > b$, which is the desired conclusion.)

14. Prove that the rationals are Archimedean.

Order completeness and Cauchy completeness

Completeness, as defined above, is called *order* completeness, because it is formulated in terms of the order of the field. Another kind of completeness, *Cauchy* completeness, is formulated in terms of distances and is related to the familiar notion of limit.

Definition. An infinite sequence of numbers $\{x(n)\}$ from an ordered field S has **limit** b (or **converges to** b) if, given any element k of S^+, there is an integer N such that

$$|x(n) - b| < k, \quad \text{for } n > N.$$

The sequence $\{x(n)\}$ is a **Cauchy sequence** if, given any element k of S^+ there is an integer M such that

$$|x(n) - x(m)| < k, \quad \text{for } m, n > M.$$

1.4. The Completeness Axiom

The terms of a Cauchy sequence get closer and closer to each other as you move out in the sequence. In this way, the sequence is trying to converge, so to speak. The definition of Cauchy sequence, however, makes no mention of a limit, and, in fact, it is possible that a sequence be Cauchy but not have a limit in the field. This does not happen, however, in a Cauchy complete field according to the following definition.

Definition. An ordered field S is **Cauchy complete** if every Cauchy sequence in S has a limit in S.

Cauchy completeness is similar to (order) completeness in intention. Both require that S be without holes of some sort. One difference between the two is that Cauchy completeness is defined using only the concept of distance which can be defined in some contexts without using order (for example in a plane). Cauchy completeness is sometimes called metric completeness, to emphasize its connection with distance.

Problems

15. Find an example of a sequence of numbers from an ordered field of your choice that is Cauchy but does not have a limit in that field.

16. If a sequence has a limit, prove that it is a Cauchy sequence.

17. Prove that a Cauchy sequence is bounded above and below.

Order completeness implies Cauchy completeness:

Theorem 1.4.2. *A complete field is Cauchy complete.*

Lemma. *In a complete field, every bounded monotonic (i.e., increasing or decreasing) sequence converges.*

Proof of the lemma. See problem 19. □

Proof of the theorem. Let S be a complete field. Let $\{x(n)\}$ be a Cauchy sequence in S. The goal is to prove that this sequence has a limit. By problem 17, the set $\{x(n)\}$ is bounded above and below. Therefore, by (order) completeness, we can define

$$y(1) = \text{lub}\,\{x(1), x(2), x(3), \ldots\},$$

and
$$y(2) = \text{lub}\,\{x(2), x(3), x(4), \ldots\},$$
and, more generally,
$$y(n) = \text{lub}\,\{x(n), x(n+1), x(n+2), \ldots\}.$$
Note that $y(n)$ is an upper bound for all the numbers for which $y(n+1)$ is least upper bound, namely, $y(n)$ is the least upper bound of a set consisting of just one more number than the set of numbers of which $y(n+1)$ is the least upper bound. Therefore
$$y(n) \geq y(n+1).$$
The new sequence $\{y(n)\}$ is nicer than $\{x(n)\}$ because $\{y(n)\}$ is monotonically decreasing and is bounded below by glb $\{x(n)\}$. Therefore, by the lemma, $\{y(n)\}$ converges to an element b.

We conclude the proof by showing that $\{x(n)\}$ also converges to b. Thus let k be an element of S^+. Because $\{x(n)\}$ is Cauchy there is an integer N such that for $n, m > N$,
$$|x(m) - x(n)| < k/2.$$
In view of the definition of $y(n)$, this means that $|y(n) - x(n)| \leq k/2$ for $n > N$. (See problem 18). Next because $\{y(n)\}$ converges to b, we can assume that N is so large that for $n > N$
$$|y(n) - b| < k/2.$$
Now for $n > N$ we have
$$|x(n) - b| \leq |x(n) - y(n)| + |y(n) - b| < k.$$
This proves the theorem. □

Problems

18. Let G be a set of numbers bounded above. Let $y = \text{lub}\,G$. Prove that if $|x + b| < k$ for all x in G, then $|y + b| \leq k$.

    (*Hint*: Rewrite the conclusion as two inequalities without the absolute value: $-k \leq y + b \leq k$. Prove them separately.)

19. Prove the lemma.

    (*Hint*: Apply the completeness axiom to a bounded monotonic sequence. Show that the sequence converges to its greatest lower bound, if the sequence is decreasing, or least upper bound, if it is increasing.)

1.4. The Completeness Axiom

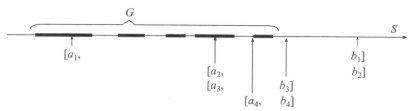

Figure 1.4.3. Proof by bisection: the first few intervals.

Completeness = Cauchy Completeness + the Archimedean property

The Archimedean property and Cauchy completeness are consequences of order completeness. Conversely, together they imply order completeness. This is useful as it allows two simpler ideas to replace the difficult concept of order completeness. The proof of this theorem is also significant. It employs a powerful technique called proof by bisection.

Theorem 1.4.3. *An ordered field that is Archimedean and Cauchy complete is order complete.*

Proof. Let S be a linearly ordered field that is both Archimedean and Cauchy complete. Let G be a non-empty subset of S that is bounded above. Our goal is to prove that G has a least upper bound.

The proof uses a trick; a trick so useful that it has become a technique: **proof by bisection**. We start with an interval $[a_1, b_1]$, which is then divided in half at its midpoint $(a_1 + b_1)/2$. One of the resulting half intervals is chosen, call it $[a_2, b_2]$, and subdivided in turn into subintervals that are quarters of the original interval. One of these subintervals, $[a_3, b_3]$, is chosen and in turn divided in half and one of those halves chosen. Continuing to choose subintervals and halve them, the i th interval obtained is called $[a_i, b_i]$. This is the setup for proof by bisection. It is illustrated in Figure 1.4.3.

What we have not yet explained is how to pick the very first interval $[a_1, b_1]$ and how to choose between the two subintervals at each step before proceeding to the next bisection. These choices depend on what one is trying to prove. However, in all proofs by bisection the left endpoints a_i increase and the right endpoints b_i decrease. In the middle is a point c that is the limit of both sequences of endpoints. Under favorable circumstances, c has whatever property is needed to complete the proof.

For the proof of Theorem 1.4.3 we choose $[a_i, b_i]$ so that b_i is an upper bound for G and a_i is not. (See Figure 1.4.3.) The proof is completed by proving:

1. it is possible to choose the subintervals $[a_i, b_i]$ as described,
2. the sequence of upper endpoints $\{b_i\}$ is a Cauchy sequence (this uses the Archimedean property of S),
3. the sequence $\{b_i\}$ converges to a limit b (this uses the Cauchy completeness of S), and, finally,
4. b is the least upper bound of G.

Proof of (1). We are given that G is a non-empty subset of S that is bounded above. Let g be an element of G (G is non-empty), let $a_1 = (g - 1)$, and let b_1 be an upper bound of G (G is bounded above) greater than a_1. Then a_1 is not an upper bound for G but b_1 is, so $[a_1, b_1]$ can be our first interval.

In order to choose the next subinterval, let $d = (a_1 + b_1)/2$. Either d is an upper bound for G or not. If d is an upper bound, then set $a_2 = a_1$, and $b_2 = d$; if d is not an upper bound, then set $a_2 = d$ and $b_2 = b_1$. In either case $[a_2, b_2]$ is a subinterval consisting of half of the first interval $[a_1, b_1]$ and such that the left endpoint a_2 is not an upper bound for G, while the right endpoint b_2 is an upper bound for G. The choices of $[a_3, b_3]$, $[a_4, b_4]$, and so on proceed in a similar manner.

Proof of (2). Let $L_1 = b_1 - a_1$ be the length of the first interval. If L_i is the length of the ith interval, then $L_i = L_1/2^i$.

Let k be an element of S^+. To prove that $\{b_i\}$ is a Cauchy sequence, we need to find an integer N such that $|b_i - b_j| < k$ for all $i, j > N$. Here it suffices to show that there is an integer N such that $L_N < k$ since the terms of the sequence $\{b_i\}$ are in the Nth subinterval from $i = N$ on.

By the Archimedean property, there is an integer N such that $L_1/N < k$. Furthermore, $2^N \geq N$ (see problem 20), so that

$$L_N = L_1/2^N \leq L_1/N < k.$$

Proof of (3). Since $\{b_i\}$ is Cauchy, it follows immediately from the Cauchy completeness of S, that the sequence converges to a limit c in S.

Proof of (4). It is clear that c is also the limit of the sequence $\{a_i\}$. Therefore given a positive integer k there are elements a_n and b_m such that

$$c - 1/k < a_n \leq c \leq b_m < c + 1/k. \qquad (*)$$

(See Figure 1.4.4.) The inequality $(*)$ can then be used to show:

1.4. The Completeness Axiom

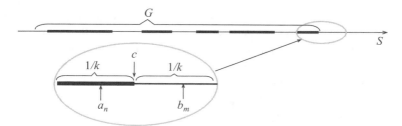

Figure 1.4.4. Proof by bisection at a later stage.

(a) that every element g of G satisfies $g \leq c$ and

(b) that every upper bound B of G satisfies $c \leq B$. In other words c is an upper bound for G, and is the least upper bound.

To prove (a) suppose, contrary to what we want to prove, that there is an element g of G such that $g > c$. By the Archimedean property there is an integer k such that $1/k < g - c$. Then, by (∗), there is an element b_m such that
$$c \leq b_m < c + 1/k < c + (g - c) = g,$$
which contradicts the fact that b_m is an upper bound for G! The proof of (b) is similar. (See Problem 21.) □

Corollary. *A field is complete if and only if it is Cauchy complete and Archimedean.*

In other words, order completeness is equivalent to metric completeness plus the Archimedean property.

Problems

20. Prove that $2^N \geq N$ for every natural number N.

    (*Hint*: Try proof by induction.)

21. Complete the proof of Theorem 1.4.3 part (4) (b).

Summary

Completeness is a technical condition whose purpose is to ensure that an ordered field have no gaps. It restricts the field in two ways that relate to properties of the field on a small scale. A complete ordered field cannot, for example, have an infinitely small element (the Archimedean property), and if the elements of a sequence from a complete ordered field are getting closer and closer to each other, then there is an element of the field toward which the sequence converges.

Completeness completes the list of axioms for the reals. In the next chapter, we prove that the reals actually exist, by constructing them, and prove that our axiom system is categorical.

2

Construction of the Reals

Why construct the reals?

We will prove the existence of the real numbers two times—by twice constructing a complete, linearly ordered field. Afterwards, we prove that all complete, linearly ordered fields are isomorphic, meaning that our axiom system for the reals is categorical.

The reader might well object to this chapter in the following terms: "I've used the reals all my life; their properties are familiar; I know how to calculate with them; I know the calculus; I know everything. Why should I bother to construct the reals, when I know the result in advance?"

How do we know the reals exist? Just because we've used a few of them over the years and assumed some properties for them does not allow us to conclude that they are really out there. Physicists argued for years about properties of the æther before it was discovered that it doesn't exist. To avoid a similar fate the reals' existence must be proved. The simplest and most reliable way to do this is to construct them.

It was empirical evidence (the Michelson-Morley experiment, 1887) that did the æther in. The reader might argue that generations of mathematicians, scientists, and engineers have used the reals without encountering any problems: doesn't that constitute experimental evidence for the existence of the reals? Certainly it does, but the goal of mathematics is to construct theories backed by stronger evidence than experiment. Experimental evidence only makes the existence of the reals plausible. It is possible, despite years of use, that the axioms of the reals contain a hidden inconsistency. By carry-

35

ing out a construction, we prove that the axioms of the reals are consistent, or, more precisely, are at least as consistent as the axioms of set theory.

Furthermore, by constructing the reals we learn more about them and we learn methods that apply to the construction of other number systems. This is the primary reason for constructing the reals here.

OK, but why construct the reals twice?

Well, perhaps two constructions is overkill. Certainly, reading this book for the first time, one may skim over (or even entirely omit) one construction. However each construction furnishes ideas used later to construct alternative number systems.

One must read at least one of the constructions to understand why the axioms of the reals are categorical. That proof depends on a construction of the reals (though it doesn't matter which).

2.1 Cantor's Construction

Real numbers are sequences of rationals

Any construction of the reals from the rationals must fill in the gaps in the rational number system. One well-known way to get at these gaps is through approximation. For example π, a fairly famous irrational, is approximated more and more closely by the familiar sequence of fractions:

$$3, \frac{31}{10}, \frac{314}{100}, \frac{3141}{1000}, \frac{31415}{10000}, \frac{314159}{100000}, \cdots,$$

derived from its decimal expansion. In Cantor's construction of the reals (first published in 1883) a sequence like this is treated as a single number. Cantor's idea is to define the reals as the set of all approximating sequences of rationals, and to manipulate sequences as though they were numbers. This means defining algebraic operations and a linear ordering for sequences.

One difficulty the reader will spot right away is that many sequences approximate the same number. For example also approaching π is the rational sequence

$$3, \frac{22}{7}, \frac{333}{106}, \frac{355}{113}, \frac{103993}{33102}, \cdots,$$

derived from its continued fraction expansion. To deal with this, we impose an equivalence relation on the set of sequences. The equivalence classes actually form the real number system.

2.1. Cantor's Construction

Two types of rational sequences are needed to carry out Cantor's ideas in detail: Cauchy sequences (already introduced in Chapter 1), and null sequences (defined below).

Definition. A **null** sequence is a convergent sequence whose limit is zero.

The next theorem gives the most important properties of these sequences.

Theorem 2.1.1. *Let $x = \{x(n)\}$ and $y = \{y(n)\}$ be sequences of rationals.*

1) *If x and y are Cauchy sequences, then so are $\{x(n) + y(n)\}$ and $\{x(n)y(n)\}$.*
2) *If x and y are null sequences, then so are $\{x(n)+y(n)\}$ and $\{x(n)y(n)\}$.*
3) *If x is a Cauchy sequence and y is a null sequence, then $\{x(n)y(n)\}$ is a null sequence.*

Proof of (1): To prove that $\{x(n) + y(n)\}$ is a Cauchy sequence, let k be a positive rational number. We must find M so that

$$|(x(n) + y(n)) - (x(m) + y(m))| < k, \text{ for } m, n > M.$$

In other words, we must make $|(x(n) + y(n)) - (x(m) + y(m))|$ small (that is, less than k) as m and n get large (i.e., become greater than M).

Well, using the triangle inequality for absolute values

$$|(x(n) + y(n)) - (x(m) + y(m))| = |x(n) - x(m) + y(n) - y(m)|$$
$$\leq |x(n) - x(m)| + |y(n) - y(m)|.$$

The advantage of this is that if the two quantities $|x(n) - x(m)|$ and $|y(n) - y(m)|$ are separately made small (say less than $k/2$) then their sum will be small and, by the above inequality,

$$|(x(n) + y(n)) - (x(m) + y(m))|$$

will also be small. This is what we want.

Now $|x(n) - x(m)|$ and $|y(n) - y(m)|$ can be made small separately because, by hypothesis, x and y are Cauchy sequences.

Here is how the proof is completed: Given $k, k/2$ is also a positive rational. Therefore, by the definition of Cauchy sequence, there are integers M_1 and M_2 such that

$$|x(n) - x(m)| < k/2 \text{ for } m, n > M_1,$$

and
$$|y(n) - y(m)| < k/2 \text{ for } m, n > M_2.$$

Let M be the larger of the two integers M_1 and M_2. Then for $m, n > M$, we have

$$\begin{aligned}|(x(n) + y(n)) - (x(m) + y(m))| &= |x(n) - x(m) + y(n) - y(m)| \\ &\leq |x(n) - x(m)| + |y(n) - y(m)| \\ &< \frac{k}{2} + \frac{k}{2} = k.\end{aligned}$$

This proves that $\{x(n) + y(n)\}$ is a Cauchy sequence.

For the product sequence $xy = \{x(n)y(n)\}$ the trick of adding and subtracting the same term must be used before the triangle inequality can be applied:

$$\begin{aligned}&|x(n)y(n) - x(m)y(m)| \\ &= |x(n)y(n) - x(n)y(m) + x(n)y(m) - x(m)y(m)| \\ &\leq |x(n)y(n) - x(n)y(m)| + |x(n)y(m) - x(m)y(m)| \\ &= |x(n)||y(n) - y(m)| + |x(n) - x(m)||y(m)|.\end{aligned}$$

Thus we can make $|x(n)y(n) - x(m)y(m)|$ small if we can make $|x(n) - x(m)|$ and $|y(n) - y(m)|$ both small (as before), and if $|x(n)|$ and $|y(m)|$ aren't large. Now, in fact, $|x(n)|$ and $|y(m)|$ are bounded above and below (according to problem 11 in 1.4), so this strategy will succeed!

Here is how the proof is completed: Let B be an integer bounding x and y, that is, such that $|x(n)| < B$ and $|y(n)| < B$. Given the positive rational k, $k/2B$ is also positive, so by the definition of Cauchy sequence there are integers M_1 and M_2 such that

$$|x(n) - x(m)| < \frac{k}{2B} \text{ for } m, n > M_1$$

and

$$|y(n) - y(m)| < \frac{k}{2B} \text{ for } m, n > M_2.$$

2.1. Cantor's Construction

Let M be the larger of M_1 and M_2. Then for $m, n > M$, we have

$$|x(n)y(n) - x(m)y(m)| \leq |x(n)|\,|y(n) - y(m)| + |x(n) - x(m)|\,|y(m)|$$
$$< B \cdot \frac{k}{2B} + \frac{k}{2B} \cdot B$$
$$= k.$$

This proves that $\{x(n)y(n)\}$ is a Cauchy sequence. □

Problems

1) Prove Theorem 2.1.1 part (2).
2) Prove Theorem 2.1.1 part (3).

Proof by trick?

Proofs often use devices so surprising that they deserve to be called tricks. But a trick used a second time becomes a technique. Every branch, sub-branch and sub-sub-branch of mathematics has its characteristic tricks. Upon repeated use they become techniques. While working with the real numbers we have already seen two such tricks/techniques: adding and subtracting something and proof by bisection.

Equivalence of Cauchy sequences

Here is the equivalence relation converting the set of Cauchy sequences into the real numbers.

Definition. Two sequences $x = \{x(n)\}$ and $y = \{y(n)\}$ are **equivalent**, written $x \sim y$, if the difference $\{x(n) - y(n)\}$ is a null sequence.

Theorem 2.1.2. *Equivalence of sequences is an equivalence relation.*

This leads to Cantor's definition of the real numbers:

Definition. Let S be the set of all Cauchy sequences of rationals. The set of **real numbers** $\mathbb{R}$ is the set of equivalence classes, $S(\backslash)$.

Note that the rational numbers $\mathbb{Q}$ are embedded in $\mathbb{R}$ as the constant sequences (e.g., the rational number $1/2$ is represented in $\mathbb{R}$ by $\{1/2, 1/2, 1/2, \ldots\}$.

Problems

3) Which of these sequences are equivalent?

$\{1/n\}$ $\{1/n^2\}$ $\{n^2/(n^2+2)\}$ $\{(-1)^n/n\}$
$\{1\}$ $\{(n^4-4)/(n^4+4)\}$ $\{0\}$

4) Let $\{x(n)\}$ be a sequence converging to a (rational) limit b. Let **x** be the equivalence class containing $\{x(n)\}$. Let $\{y(n)\}$ be another sequence in the equivalence class **x**. Prove that $\{y(n)\}$ converges to b.

5) Prove Theorem 2.1.2.

The algebra of the Cantor reals

Our goal is to prove that $\mathbb{R}$, as defined by Cantor, satisfies all the axioms of the reals listed in Chapter 1. For the field axioms, we first define addition and multiplication:

Definition. Let **x** and **y** be in $\mathbb{R}$. Let $\{x(n)\}$ be a sequence belonging to **x**, and let $\{y(n)\}$ be a sequence belonging to **y**. Addition and multiplication for $\mathbb{R}$ are defined by saying that $\mathbf{x} + \mathbf{y}$ is the equivalence class that contains the sequence $\{x(n) + y(n)\}$, and **xy** is the equivalence class that contains $\{x(n)y(n)\}$.

Whenever an operation is defined on a set of equivalence classes, one must verify that it is well-defined. This is stated in the following theorem.

Theorem 2.1.3. *If $\{x(n)\}$ and $\{x'(n)\}$ are equivalent Cauchy sequences and likewise $\{y(n)\}$ and $\{y'(n)\}$, then $\{x(n) + y(n)\}$ and $\{x'(n) + y'(n)\}$ are equivalent, and $\{x(n)y(n)\}$ and $\{x'(n)y'(n)\}$ are equivalent.*

Loosely speaking Theorem 2.1.3 says: equivalent sequences added to equivalent sequences are equivalent, and equivalent sequences multiplied by equivalent sequences are equivalent.

2.1. Cantor's Construction

Theorem 2.1.4. $\mathbb{R}$ *is a field.*

Problems

6) Prove that addition and multiplication of the Cantor reals is well-defined.

   (*Hint*: Use the results on null sequences proved earlier.)

7) Prove this lemma: If **x** is not zero, and $\{x(n)\}$ is a sequence from **x**, then there are positive integers M and N so that when $n > N$, then $|x(n)| > 1/M$.

   (*Hint*: Use the Cauchyness of the sequence $\{x(n)\}$.)

8) Prove Theorem 2.1.4.

   (*Hint*: All the field axioms must be verified. Of greatest interest and complexity is the proof that if **x** is not zero, then there is a multiplicative inverse for **x**. For this use the preceding problem.)

The geometry of the Cantor reals

We turn to ordering the Cantor reals. This requires the concept of a positive sequence and positive equivalence class of sequences:

Definition. A Cauchy sequence of rationals $\{x(n)\}$ is called **positive** if there exist positive integers M and N so that if $n > N$, then $x(n) > 1/M$.

If **x** is in $\mathbb{R}$, we say that **x** is **positive** if one of the sequences in **x** is positive.

Is the concept of positive real well-defined? The next theorem resolves this.

Theorem 2.1.5. *Let* **x** *be a real number. If one sequence from* **x** *is positive, then all sequences in* **x** *are positive.*

Now we have

Theorem 2.1.6. $\mathbb{R}$ *is an ordered field.*

Problems

9) Prove Theorem 2.1.5.

10) Prove that the positive Cantor reals are closed under addition and multiplication. This is the first part of the proof of Theorem 2.1.6.

11) Prove that the Cantor reals satisfy the trichotomy law. This completes the proof of Theorem 2.1.6.

    (*Hint*: There are two parts to the proof. The first is to show that the three conditions: **x** is positive, $-\mathbf{x}$ is positive, and $\mathbf{x} = 0$ are mutually exclusive. The second part is to show that every equivalence class **x** in $\mathbb{R}$ satisfies one of them. Perhaps the simplest way to attack the second part is to prove that if **x** is not positive and $-\mathbf{x}$ is not positive, then $\mathbf{x} = 0$. Use Cauchyness to prove that any sequence $\{x(n)\}$ in **x** is a null sequence.)

Completeness of the Cantor reals

Since it is completeness that sets the reals apart from other number systems, the next theorem is the climax of Cantor's construction.

Theorem 2.1.7. $\mathbb{R}$ *is complete.*

Problems

12) Prove that the Cantor reals are Archimedean.

    (*Hint*: Let **x** and **y** be positive reals. Show that there exist rational numbers a and b such that $0 < a < \mathbf{x}$ and $\mathbf{y} < b$. Using the Archimedean property of the rationals, find an integer n such that $na > b$. It then follows that $n\mathbf{x} > \mathbf{y}$.)

13) Prove that the Cantor reals are Cauchy complete.

    (*Hint*: To prove that $\mathbb{R}$ is Cauchy complete, let $\{\mathbf{x}(n)\}$ be a Cauchy sequence in $\mathbb{R}$. That is, $\mathbf{x}(n)$ is an equivalence class of Cauchy sequences for each n. Show that there exists a rational $b(n)$ such that $|\mathbf{x}(n) - b(n)| < 1/n$.)

    (*Detailed Hint*: Show that there exists a double sequence $\{x(n, m)\}$ of rationals such that the sequence $\{x(n, m)\}$ (holding n constant, but varying m) represents the real $\mathbf{x}(n)$. Use this double sequence to prove that the sequence $\{b(n)\}$ is a Cauchy sequence of rationals. Conclude that $\{b(n)\}$ defines an element $\mathbf{b}$ of $\mathbb{R}$. Finally, prove that $\mathbf{b}$ is the limit of the sequence $\{\mathbf{x}(n)\}$.)

14) Combine problems 12 and 13 to prove Theorem 2.1.7.

Summary

Cantor conceived of the reals as equivalence classes of sequences. To implement this idea requires defining addition, multiplication, and order for sequences. Then the axioms of the reals can be verified, completing Cantor's construction. A similar construction will be used to build the constructive reals (Chapter 5) and the hyperreals (Chapter 6).

2.2 Dedekind's Construction of the Reals

Dedekind cuts

Dedekind has the distinction of being the first to construct the reals (in the mid 1800s). His construction differs substantially from Cantor's and is not as straightforward. The two make an interesting pair. With Cantor's construction, the algebraic and geometric axioms are easy to verify, while the proof of completeness is long. With Dedekind's construction, completeness is easy, but some of the algebraic details require lengthy verification.

Definition. A **Dedekind cut** (or simply a **cut**) is a subset $\mathbf{x}$ of the rationals $\mathbb{Q}$ such that

(a) neither $\mathbf{x}$ nor the complement of $\mathbf{x}$ is empty,

(b) if r is in $\mathbf{x}$, and $s > r$, then s is in $\mathbf{x}$,

(c) $\mathbf{x}$ has no smallest element.

With this definition, the Dedekind reals are easily defined.

Definition. The set of **real numbers** $\mathbb{R}$ is defined as the set of all cuts.

One example of a cut is the set of positive rationals; others are given in the problems.

Problems

1) Prove that the positive rationals, $\mathbb{Q}+$, is a cut (called the **null** cut).

2) If x is a rational number, verify that $\{r \mid r > x\}$ is a cut. These cuts are called **rational** cuts.

3) Prove that the set $\{r \mid r > 0, r^2 > 2\}$ is a cut.

4) Prove that the union of a finite number of cuts is a cut.

Addition of the Dedekind reals

This is easy to define.

Definition. If **x** and **y** are cuts, then the sum $\mathbf{x} + \mathbf{y}$ is defined.

$$\mathbf{x} + \mathbf{y} = \{r + s \mid r \in \mathbf{x} \text{ and } s \in \mathbf{y}\}.$$

Theorem 2.2.1. *Addition of cuts is well-defined, that is, $\mathbf{x} + \mathbf{y}$ is a cut.*

Theorem 2.2.2. *Addition on $\mathbb{R}$ satisfies the laws of commutativity, associativity, additive identity, and additive inverse.*

Problems

5) Prove that addition of cuts is well-defined.

6) Prove Theorem 2.2.2.

   (*Hint*: Most parts of this proof are straightforward. The last part requires a definition: the **negative** of a cut **x** is

   $$-\mathbf{x} = \{-r \mid r \text{ is neither in } \mathbf{x} \text{ nor is the glb of } \mathbf{x}\}.$$

   Note: The glb of **x** is specifically excluded so that $-\mathbf{x}$ will satisfy property (c) of cuts.)

2.2. Dedekind's Construction of the Reals

Geometry of the Dedekind reals

The multiplication of Dedekind cuts is not simple. The problem is determining the sign of a product. It is best to start with multiplication of the *positive* Dedekind reals, but then we must define positive cuts:

Definition. A cut is **non-negative** if it is a subset of the null cut. It is called **positive** if it is a proper subset of the null cut.

Theorem 2.2.3. *The positive cuts are closed under addition. The Dedekind reals satisfy the trichotomy law.*

Theorem 2.2.4. $\mathbb{R}$ *is complete.*

Thus, except for properties of multiplication, we now know that the Dedekind reals form a complete ordered field.

Problems

7) Prove Theorem 2.2.3.

8) Let **x** and **y** be cuts. Prove that $\mathbf{x} < \mathbf{y}$ if and only if $\mathbf{x} \supseteq \mathbf{y}$.

9) Prove Theorem 2.2.4.

   (*Hint*: Let G be a subset of $\mathbb{R}$ that is bounded above. The problem is to prove that G has a least upper bound. Form the union of all cuts that are upper bounds for G. Use problem 8 to show that the union is a cut.)

Multiplication of Dedekind Reals

For non-negative reals multiplication is simple.

Definition. If **x** and **y** are non-negative cuts, their product is defined by

$$\mathbf{xy} = \{rs | r \in \mathbf{x} \text{ and } s \in \mathbf{y}\}.$$

Theorem 2.2.5. *Multiplication is well-defined for non-negative cuts and the set of non-negative cuts is closed under multiplication. In addition, multiplication satisfies the laws of commutativity, associativity, identity, inverse, and the distributive law (non-negative cuts only).*

Problem
---
9) Prove Theorem 2.2.5.

This completes the construction of the Dedekind reals except for the extension of multiplication to non-negative reals. The theory at this point becomes complicated due to the difficulty of determining the signs of products. For completeness we state the following definition.

Definition. If **x** is a cut, exactly one of **x** and −**x** is non-negative. Thus we define:

$$|\mathbf{x}| = \text{the non-negative one of } \mathbf{x} \text{ and } -\mathbf{x}.$$

For cuts **x** and **y** we define

$$\mathbf{xy} = \begin{cases} |\mathbf{x}||\mathbf{y}| & \text{if } \mathbf{x} \text{ and } \mathbf{y} \text{ are both positive,} \\ |\mathbf{x}||\mathbf{y}| & \text{if } \mathbf{x} \text{ and } \mathbf{y} \text{ are both negative,} \\ -|\mathbf{x}||\mathbf{y}| & \text{otherwise.} \end{cases}$$

Theorem 2.2.6. *With this definition of multiplication, $\mathbb{R}$ is a complete, ordered field.*

The proof of this theorem is tedious since the verification of each axiom breaks down into many separate cases depending on the signs of the different quantities.

This completes Dedekind's construction of the reals. We use an analogous construction to build the surreals (in Chapter 7).

2.3 Uniqueness of the Reals

So, just what are the real numbers?

The reader might want to know: What are the reals *really*? Are they equivalence classes of Cauchy sequences? Are they Dedekind cuts? Or what?

Without entirely settling the matter, we prove that it doesn't matter what the reals are: all complete, ordered fields are identical. More precisely, given two complete, ordered fields (that is, two versions of the reals), there is a one-to-one correspondence between them that is both a field and an order

2.3. Uniqueness of the Reals

isomorphism, that is, they have the same algebra and geometry. This proves that the axioms for the reals are categorical.

That all versions of the reals are algebraically and geometrically isomorphic does not necessarily answer the question: what is a real number? Some readers will be dissatisfied that different constructions result in such different kinds of entities. To them a number like $\sqrt{2}$, for example, should be a definite thing, and not an equivalence class of Cauchy sequences or a Dedekind cut. For some the nature of the real numbers is not settled by these constructions; it remains a problem in the philosophy of mathematics.

That all complete, ordered fields are isomorphic does clarify, however, why the real numbers are so important, so fundamental. They are the only numbers that obey the basic rules of algebra (the field axioms), have the geometry of a straight line (the order axioms), and are without gaps (completeness). We proceed to prove this.

The Rational Subfield

Let S be a complete, ordered field. Our goal is to prove that S is isomorphic to $\mathbb{R}$. In this section we start by showing that S contains a subfield isomorphic to the rational numbers.

Definition. Let 1 be the multiplicative identity of S. For every integer n, define

$$i(n) = \begin{cases} 1 + 1 + \cdots + 1 & \text{— } n \text{ ones, if } n < 0 \\ -1 - 1 - \cdots - 1 & \text{— } n \text{ minus ones, if } n < 0. \end{cases}$$

For a rational number p/q define

$$i(p/q) = i(p)/i(q).$$

This allows us to extend the embedding theorems of Chapter 1 to the rationals.

Theorem 2.3.1. *The function i is well-defined. It is a field and order isomorphism from the set of rationals $\mathbb{Q}$ onto a subfield of S.*

This theorem says that S contains a subfield that is in every algebraic and geometrical sense identical to the rationals. From now on we simply assume that the field S contains the rationals. The next theorem explains how the rationals are distributed inside S.

Theorem 2.3.2. *Let a and b be two elements of S and assume that $a < b$. Then there is a rational number q between a and b, that is, there are integers m and n, such that if $q = m/n$ then $a < q < b$.*

The conclusion of this theorem is summarized by saying the rationals are **dense** in S. The denseness of $\mathbb{Q}$ in S implies, as we see in the proof of the next theorem, that the rational subfield (together with the operations of lub and glb) determines the whole field S.

Problems

1) Prove that i as defined above is well-defined, that is that $i(p/q) = i(r/s)$ when $p/q = r/s$.

2) Prove Theorem 2.3.1.

3) Prove Theorem 2.3.2.

   (*Hint*: Choose n so that $0 < 1/n < ba$. (Why is this possible?). Let $m = \lceil na \rceil + 1$. (See Problem 11 in Section 1.4.) Let $q = m/n$.)

Isomorphism of Complete Ordered Fields

The following theorem completes the proof of the uniqueness of the reals.

Theorem 2.3.3. *Every complete, ordered field is isomorphic to $\mathbb{R}$.*

Proof. Let S be a complete, ordered field. We must construct an isomorphism $i : \mathbb{R} \to S$. We have just seen that S contains a dense copy of the rational numbers $\mathbb{Q}$. The isomorphism i can begin by mapping each rational in $\mathbb{R}$ to the corresponding rational in S. The problem is to extend i from the rationals to the rest of $\mathbb{R}$. At this point, we choose $\mathbb{R}$ to be either the Cantor or Dedekind reals. The rest of the proof varies in detail depending on this choice (see problems 4 and 5). □

Corollary. *The Cantor reals and the Dedekind reals are isomorphic.*

2.3. Uniqueness of the Reals

Problems
---

4) Complete the proof of Theorem 2.3.3 using the Cantor reals.

   (*Hint*: For $\mathbb{R}$ the Cantor reals, proceed as follows: each **x** in $\mathbb{R}$ is an equivalence class of Cauchy sequences of rationals. If $\{x(n)\}$ is a representative sequence for **x**, define

   $$i(\mathbf{x}) = \lim_{n \to \infty} i(x(n)),$$

   where the limit is taken in S. The limit exists because S is complete and because $i(x(n))$ is a Cauchy sequence in S. (Why?)

   Prove: (1) i is well-defined, (2) i is one-to-one, (3) i maps $\mathbb{R}$ onto S, (4) i is a field isomorphism, and (5) i is an order isomorphism.

5) Complete the proof of Theorem 2.3.3 using the Dedekind reals.

   (*Hint*: For $\mathbb{R}$ the Dedekind reals, proceed as follows: each **x** in $\mathbb{R}$ is cut in the rationals. Therefore we can define

   $$i(\mathbf{x}) = \mathrm{glb}\{i(r) | r \in \mathbf{x}\},$$

   where the greatest lower bound is taken in S. The greatest lower bound exists because S is complete and because $\{i(r) | r \in \mathbf{x}\}$ is a set with a lower bound in S. (Why?)

   Prove: (1) i is well-defined, (2) i is one-to-one, (3) i maps $\mathbb{R}$ onto S, (4) i is a field isomorphism, and (5) i is an order isomorphism.

6) Let S be an ordered field. A gap in S is a pair $\{X, Y\}$ of subsets of S such that:

   (a) $X \cap Y = \emptyset$ and $X \cap Y = S$,
   (b) if $x \in X$ and $y \in Y$, then $x < y$,
   (c) X has no greatest element and Y has no least element.

   Prove that if S has no gaps, then S is field and order isomorphic to $\mathbb{R}$.

Summary

This section completes the agenda announced at the beginning of Chapter 1: the reals have been constructed and their fundamental properties have proven to be categorical.

2.4 The Differential Calculus

The construction of the reals is the start of a long story. Whole branches of mathematics including much of the calculus are, in a sense, parts of this story. We can't begin to cover all this here. Instead we list a few important theorems that extend the process of construction which we have just finished. It will be amusing to track how these results alter as we move from the reals through various alternative number systems.

Continuous Functions

Continuity is fundamental to the calculus. It is a simple form of smoothness, yet working with it is far from easy. As examples, consider the intermediate value theorem, the boundedness theorem and the maximum value theorem (statements of these follow). These are important results about continuous functions that seem intuitively obvious, yet none are easy to prove. Significantly, their proofs depend on the completeness of the real numbers, which we have gone to such pains to develop.

In this section we outline proofs of these fundamental results in the interest of examining how these results come out in other number systems.

First the definition of continuity.

Definition. Let f be a function defined on the interval $[a, b]$. Then f is called **continuous** on $[a, b]$ if for any sequence $\{x(n)\}$ in $[a, b]$ with limit k,

$$\lim_{n \to \infty} f(x(n)) = f(k).$$

Intuitively a continuous function is one whose graph can be drawn without taking your pencil off the paper. The intermediate value theorem, one of the most famous and important theorems about continuity, expresses this intuitive idea precisely.

Theorem 2.4.1. *(Intermediate value theorem for continuous functions.) Let f be a continuous function on the closed interval $[a, b]$. If $f(a) < 0$ and $f(b) > 0$, then there is a real number c, $a < c < b$, such that $f(c) = 0$.*

Problems

1) Prove the intermediate value theorem.

2.4. The Differential Calculus

(*Hint*: Use proof by bisection. Choose successive subintervals $[a_k, b_k]$ of the interval $[a, b]$ so that $f(a_k) < 0$ and $f(b_k) > 0$. Let c be the limit of a_k.)

2) Use the intermediate value theorem to prove that every positive real number has a positive square root, that is, for every real $x > 0$, there is a real $y > 0$ such that $y^2 = x$.

3) Prove that f is continuous on $[a, b]$ if and only if for every $\varepsilon > 0$ and c in $[a, b]$ there is a $\delta > 0$ such that if $|x - c| \leq \varepsilon$ then $|f(x) - f(c)| \leq \delta$.

4) Prove the boundedness theorem for continuous functions: If f is a continuous function on a closed interval $[a, b]$ then there is a constant K such that $|f(x)| < K$, for all x in $[a, b]$.

(*Hint*: Suppose that f is not bounded. Use proof by bisection to deduce that there must be a point c in $[a, b]$ at which f has limit $\pm\infty$. This contradicts the fact that f has a finite value at c.

5) Prove the maximum value theorem for continuous functions. If f is a continuous function on a closed interval $[a, b]$ then there is a point c in $[a, b]$ such that $f(x) \leq f(c)$ for all x in $[a, b]$. In other words, $f(c)$ is a maximum value of the function f on the interval $[a, b]$.

(*Hint*: Use proof by bisection.)

The Derivative

The differential calculus is based on this definition.

Definition. Let the function $f(x)$ be defined on the interval $[a, b]$. Let x be a point in $[a, b]$. Then f is **differentiable** at x with derivative $f'(x)$ if the following limit exists:

$$f'(x) = \lim_{h \to 0} \frac{f(x + h) - f(x)}{h}.$$

We make no pretense of developing the calculus in any detail. However, it will be interesting to take one or two typical results and observe how they must be adapted to the number systems presented in subsequent chapters.

Problems

6) Prove the product rule:
$$(f(x)g(x))' = f'(x)g(x) + f(x)g'(x).$$

(*Hint*: Add and subtract something.)

7) Prove the chain rule:
$$f(g(x))' = f'(g(x))g'(x).$$

(*Hint*: The chain rule is more difficult conceptually than the product rule, and its proof is trickier. A natural proof starts with:
$$\lim_{x \to c} \frac{f(g(x)) - f(g(c))}{x - c} = \lim_{x \to c} \frac{f(g(x)) - f(g(c))}{g(x) - g(c)} \frac{g(x) - g(c)}{x - c},$$
which appears to give the desired result since as x approaches c, $g(x)$ will approach $g(c)$ (differentiable functions are continuous), so that the two factors on the right tend to $f'(g(x))$ and $g'(x)$ respectively. This simple computation is the reason why the chain rule works, however a technical problem prevents this from being a correct proof: the denominator $g(x) - g(c)$ may be zero. A proof must be found that avoids dividing by zero!)

2.5 A Final Word about the Reals

The reals go back to the most ancient of periods in the history of mathematics, at least in the form of natural numbers and simple fractions. The first number known to be irrational was the square root of five, which is connected with the geometry of a regular pentagon. Its irrationality was discovered by the Pythagoreans. According to legend, this discovery caused a crisis since the brotherhood of Pythagoreans believed in the primacy of integers: specifically, although they would not have put it this way, they held that all numbers could be expressed by finite combinations of integers.

Considering their roots in antiquity, it is perhaps surprising that the theoretical foundation of the real numbers should be of such recent formulation. The work of Richard Dedekind (1831–1916) on the construction of the reals

2.5. A Final Word about the Reals

occurred in the period 1852–1872, the last year being the date of publication of his theory. The idea of Cauchy sequences was introduced by Augustin-Louis Cauchy (17890–1857) in 1821 in connection with his development of the calculus, but was not used to describe the real numbers until Georg Cantor (1845–1918) took up this problem, publishing his theory of the reals in 1883.

The development of the theory of the real numbers was only part of a large scale mathematical movement spanning the whole nineteenth century whose goal was to construct a satisfactory logical basis for the calculus. A remarkable thing about this movement is that it worked its way backwards through the number systems. First to be treated were the fundamental theorems of the calculus itself, whose proofs can be based directly on the idea of limit (for example, the intermediate value theorem). As this theory reached a satisfactory state, it was realized that these proofs depended on several fundamental assumptions about limits that in turn needed proof. Typical of these assumptions is the statement that every bounded monotonic sequence converges. Most of these assumptions turned out to involve the completeness of the reals. This realization led to the development of the theory of the real numbers described in these notes. Finally attention turned (even more fundamentally) to the integers and rational numbers whose theory was given definitive formulation by Giuseppe Peano (1858–1932) in 1889. This movement through the number systems was systematic, methodical, and logical, although seemingly backwards.

Work on the foundations of the reals and other number systems has led to new results and new areas of research still being pursued today. In particular, clarification of the foundation of the real numbers led naturally to the development of such fields as topology, set theory, abstract algebra, mathematical logic, and differential geometry, all of which were of major importance in twentieth-century mathematics.

References: This chapter is based on [E2] and [E3]. Two excellent comprehensive sources on number systems are [A1] and [A2], while [A3] is an outstanding resource on the history of mathematics. Readers interested in a detailed discussion of how to find proofs can consult references [B1] through [B5].

Part II

MULTI-DIMENSIONAL NUMBERS

The real numbers are the only complete, linearly ordered field. Any other system of numbers we construct will either be incomplete, not linearly ordered, or not a field. Consider these three possibilities briefly.

The most fundamental aspect of any system of numbers is their algebra. Our intention is to consider number systems that can seriously be used in place of the reals. So number systems that are not fields, stray too far, we believe, from what makes the reals the reals. Thus, we avoid systems that are not fields (with one exception).

Linear order is another matter. There are many practical problems to which the geometry of a line does not apply. Therefore, we shall consider number systems that are complete fields, but not linearly ordered. To simplify this situation, we only consider systems with the geometry of n-dimensional space (for some integer n). With this restriction, we give a complete account of all such systems.

Finally, there are several interesting number systems that preserve the algebra and the geometry of the reals, but are incomplete. They appear in Part III.

3

The Complex Numbers

3.1 Two-Dimensional Algebra and Geometry

Introducing the complexes

The complex numbers are the oldest and best-known extension of the reals. They have an elaborate theory and many important applications. Complex numbers were viewed with suspicion by mathematicians for many years and used only with reluctance because they seemed to have no basis in reality. True understanding of the complexes came after it was grasped that they are a two-dimensional number system. Interpreting the complexes as points in the plane gave them concreteness and opened the door to their exploration and application.

Algebra of the complex numbers

For convenience, we define the complexes as certain 2 by 2 matrices. This gives the theory a head start since matrices already have a well-known algebraic structure:

Definition. A **complex number** is a (2 by 2) real matrix of the form

$$\begin{bmatrix} a & b \\ -b & a \end{bmatrix}.$$

Addition and multiplication of complex numbers are the usual matrix operations, namely

$$\begin{bmatrix} a & b \\ -b & a \end{bmatrix} + \begin{bmatrix} c & d \\ -d & c \end{bmatrix} = \begin{bmatrix} a+c & b+d \\ -(b+d) & a+c \end{bmatrix},$$

and

$$\begin{bmatrix} a & b \\ -b & a \end{bmatrix} \begin{bmatrix} c & d \\ -d & c \end{bmatrix} = \begin{bmatrix} ac-bd & ad+bc \\ -(bc+ad) & -bd+ac \end{bmatrix}.$$

The set of all complex numbers is written $\mathbb{C}$.

Although we have defined the complexes as 2 by 2 matrices, which are four-dimensional in nature, the complex numbers are two-dimensional. To make this clear we introduce the matrix

$$i = \begin{bmatrix} 0 & 1 \\ -1 & 0 \end{bmatrix}.$$

Then every complex number can be written in the form $z = a + bi$ where the matrix

$$\begin{bmatrix} a & 0 \\ 0 & a \end{bmatrix}$$

is abbreviated a. This is the **Cartesian form** of the complex number. The quantity a is the **real** part of z and written $\mathrm{Re}(z)$, while b is the **imaginary** pa, written $\mathrm{Im}(z)$. Both the real and imaginary parts of z are real.

Theorem 3.1.1. $\mathbb{C}$ *is a field.*

Problems

1) Verify that i is a square root of -1.

2) Prove Theorem 3.1.1

   (*Hint*: The algebra of matrices already has most of the properties of a field. Prove that the complex numbers are closed under the operations of matrix addition and multiplication.)

3) Prove that i and $-i$ are the only square roots of -1.

4) Express these complex numbers in Cartesian form:

   (a) $(1 + 2i)(3 - 4i)$

3.1. Two-Dimensional Algebra and Geometry

    (b) $(i - 1)^2$

    (c) $(i - 1)^4$

5) Express these complex numbers in Cartesian form:

    (a) $\frac{1}{1-2i}$

    (b) $\frac{1}{1+i}$

    (c) $\frac{-2}{1+i\sqrt{3}}$

    (d) $\frac{1+2i}{2+2i}$

(*Example*:
$$\frac{1}{1-i} = \frac{1}{1-i}\frac{1+i}{1+i} = \frac{1+i}{2}.$$
This is called rationalizing the denominator.)

6) Let $z = a + bi$ and $w = c + di$. Write the product zw in Cartesian form.

Geometry of the complex numbers

As a two-dimensional number system, it is almost obvious that the complex field is not linearly ordered. This is, nonetheless, important enough to state as a theorem. For a proof, see problem 7.

Theorem 3.1.2. *The complex numbers cannot be linearly ordered.*

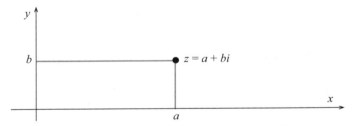

Figure 3.1.1. Complex numbers form a plane.

As two-dimensional quantities, it is natural to visualize the complex numbers as points in the Cartesian plane (as in Figure 3.1.1). The planar nature of the complex numbers is the source of many important applications. The next definition introduces some geometric quantities:

Definition. For a complex number $z = a + bi$, let

$\bar{z} = a - bi$, — the **conjugate**

$|z| = \sqrt{a^2 + b^2}$, — the **modulus**

$\arg(z)$ = the polar angle θ, — the **argument**

These are illustrated in Figure 3.1.2. Note that $|z|$ and $\theta = \arg(z)$ are the polar coordinates of the Cartesian point (a, b).

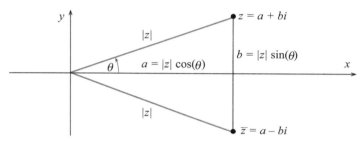

Figure 3.1.2. Locating a complex number in the plane.

Problems

7) Prove Theorem 3.1.2.

   (*Hint*: Give an indirect proof. Compare i with 0. If the trichotomy law held, then either $i > 0$, $i < 0$, or $i = 0$. Each leads to a contradiction.)

8) Verify these properties of the conjugate:
   (a) $\overline{z + w} = \bar{z} + \bar{w}$,
   (b) $\overline{zw} = \bar{z}\,\bar{w}$,
   (c) $\bar{\bar{z}} = z$,
   (d) $\bar{z}$ is the reflection of z across the x-axis,
   (e) $\mathrm{Re}(z) = \frac{z + \bar{z}}{2}$,
   (f) $\mathrm{Im}(z) = \frac{z - \bar{z}}{2i}$.

9) Verify these properties of the modulus:
   (a) $|kz| = k|z|$, for real $k > 0$,
   (b) $|z + w| \le |z| + |w|$,
   (c) $|z|^2 = z\bar{z}$,

3.1. Two-Dimensional Algebra and Geometry

(d) $|z - w|$ is the Euclidean distance from z to w,

(e) $|z| = 0$ if and only if $z = 0$,

(f) $z^{-1} = \frac{\bar{z}}{|z|^2}$.

10) Considering the complex numbers as matrices, what terms from matrix algebra are used for the conjugate and modulus?

11) Let $z = 2+i$. On the same pair of axes plot the points $z, 1/z, -z, -1/z$ and their conjugates.

12) Sketch the set of points z in the plane such that

(a) $|z - 1| = 4$,

(b) $|z + i| = 2$,

(c) $|z + 2| < 5$,

(d) $|z - 1| = |z|$,

(e) $|z + i| = |z - i|$,

(f) $|4z + i| > 2$.

Completeness of the complex numbers

$\mathbb{C}$ is not an ordered field, so order completeness makes no sense. Nor does the Archimedean property. However, because the complex numbers correspond to the points of a plane, they have a distance function (problem 9(d)). Therefore we can define limits in $\mathbb{C}$ and then ask whether $\mathbb{C}$ is Cauchy (i.e., metric) complete.

Definition. A sequence of complex numbers $z(n)$ has **limit** the complex number w, if for any integer N there is an integer M such that for $n > M$ we have

$$|z(n) - w| < 1/N.$$

A sequence of complex numbers is **Cauchy** if for any integer N there is an integer M such that for $m, n > M$ we have

$$|z(m) - z(n)| < 1/N.$$

The notions of complex limit and of Cauchy sequence are directly reducible to real limits and real Cauchy sequences according to this lemma:

Lemma. *A sequence $\{z(n)\}$ of complex numbers has a limit if and only if both real sequences $\{\text{Re}(z(n))\}$ and $\{\text{Im}(z(n))\}$ have a limit. A sequence $\{z(n)\}$ of complex numbers is Cauchy if and only if $\{\text{Re}(z(n))\}$ and $\{\text{Im}(z(n))\}$ are Cauchy.*

The lemma makes possible the proof of the following theorem:

Theorem 3.1.3. $\mathbb{C}$ *is Cauchy complete, that is, every sequence of complex numbers that is Cauchy has a limit in the complex numbers.*

Problems

13) Prove the lemma.

   (*Hint*: Prove and use these inequalities:
   $$|\text{Re}(z)| \leq |z| \leq |\text{Re}(z)| + |\text{Im}(z)|,$$
   $$|\text{Im}(z)| \leq |z| \leq |\text{Re}(z)| + |\text{Im}(z)|.)$$

14) Prove Theorem 3.1.3.

Summary

Like the reals, the complex numbers are a field; unlike the reals, they are a plane, not a line. They are Cauchy complete. The only axioms of the reals not satisfied by the complex numbers are those that concern linear order.

3.2 The Polar Form of a Complex Number

The complex exponential

It is the relationship between algebra and geometry that makes the complex numbers useful. A central tool is the polar form of a complex number. The Cartesian form of a complex number uses Cartesian coordinates; the polar form uses polar coordinates. It is important because it supplies a geometric meaning to the operation of complex multiplication.

The polar form requires the complex exponential, defined by Euler as follows.

Definition. The complex exponential is defined by
$$\exp(i\theta) = e^{i\theta} = \cos(\theta) + i\sin(\theta).$$

3.2. The Polar Form of a Complex Number

This definition is justified by the many ways that it ties together properties of real exponential functions (i.e., the laws of exponents) and properties of the trigonometric functions. For example, the fact that $e^0 = 1$ (a law of exponents) agrees with the values $\cos(0) = 1$ and $\sin(0) = 0$. More examples are in problem 1.

Problems

1) Show that
   (a) $e^{i\theta} e^{i\tau} = e^{i(\theta + \tau)}$,
   (b) $d(e^{i\theta})/d\theta = i e^{i\theta}$,
   (c) $(e^{i\theta})^n = e^{in\theta}$.

   (*Note*: Each of these equations expresses a law of real exponents extended to complex exponents. Euler's definition converts each of them into an equation expressing properties of the trigonometric functions.)

2) Plot the set of points $\{e^{i\theta} | 0 \leq \theta \leq 2\pi\}$ in the complex plane.

Polar form and complex multiplication

For a complex number $z = a + bi$, we write

$$z = r\cos(\theta) + ir\sin(\theta)$$
$$= r(\cos(\theta) + i\sin(\theta))$$
$$= |z|e^{i\theta},$$

where $\theta = \arg(z)$ and $r = |z|$. This is the **polar form** of z (see Figure 3.2.1).

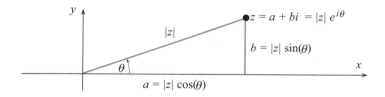

Figure 3.2.1. Polar form.

3. The Complex Numbers

If we multiply two complex numbers in polar form: $z = re^{i\theta}$ and $w = se^{i\varphi}$. We get:
$$zw = re^{i\theta}se^{i\varphi} = rse^{i(\theta+\varphi)}.$$

Thus a simple way to multiply complex numbers is to put them in polar form, then multiply the moduli and add the arguments. In formulas,
$$|zw| = |z||w|,$$
$$\arg(zw) = \arg(z) + \arg(w).$$

This is algebraically simple and makes geometric sense of complex multiplication (see Figure 3.2.2.).

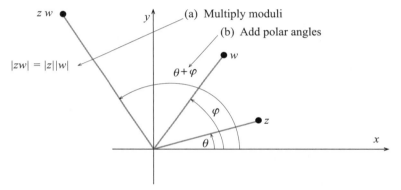

Figure 3.2.2. Multiplication of complex numbers using polar form.

Problems

3) Put into polar form

   (a) $1 + i1$
   (b) $i - \sqrt{3}$
   (c) $-i$
   (d) $-3 - 3i$
   (e) -1

4) Let $z = a + bi = |z|e^{i\theta}$ and $w = c + di = |w|e^{i\varphi}$. Prove algebraically that $|zw| = |z||w|$.

3.2. The Polar Form of a Complex Number

5) Let $w = f(z) = az$, where $a = re^{i\theta}$ is a complex constant in polar form. Show that w is obtained by stretching (or compressing) z away from (or towards) the origin by the factor r while rotating about the origin by θ.

6) Show that every complex number has two square roots. What is the geometric relationship between them?

   (*Hint*: Use polar form.)

7) Find these square roots in polar form.

   (a) $\sqrt{4i}$
   (b) $\sqrt{-\sqrt{3}+i}$
   (c) $\sqrt{1+i}$

8) Verify these formulas. (See Figure 3.2.3.)

   (a) $\text{Re}(z\bar{w}) = |z||w|\cos(\psi) = $ the dot product of z and w as vectors,
   (b) $\text{Im}(z\bar{w}) = |z||w|\sin(\psi) = $ twice the area of triangle $\triangle Ozw$.

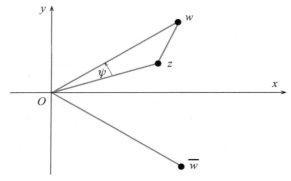

Figure 3.2.3. More complex geometry.

9) Prove that a complex number $z = re^{i\theta}$ has n n-th roots given by:

$$\sqrt[n]{z} = \sqrt[n]{r}e^{ik\theta/n},$$

for $k = 1, 2, 3, \ldots, n$. Plot them in the complex plane

   (a) for $z = 1, n = 3$
   (b) $z = 1+i, n = 4$

10) Give an appropriate definition for e^z, for $z = x + iy$, an arbitrary complex number (not purely imaginary).

   (*Hint*: Use the laws of exponents.)

11) Use your answer to problem 10 to solve these equations:

   (a) $e^z = 0$
   (b) $e^z = 1$
   (c) $e^z = 2i$

3.3 Uniqueness of the Complex Numbers

The complex numbers are a two-dimensional field. The reader may perhaps hope that we will next present a three-dimensional field, then a four-dimensional field and so on—but this is impossible. In this section we prove that the complex numbers are the only finite-dimensional field properly containing the reals. This unexpected circumstance, like the uniqueness of the reals proved in Chapter 2, sums up the significance of the complex numbers. They are the only multi-dimensional field.

We begin with a famous result:

The Fundamental Theorem of Algebra. *Every polynomial $p(z)$ of degree n with complex coefficients factors into n linear factors with complex coefficients.*

The proof involves many ideas (primarily from the theory of complex differentiation and integration) irrelevant to the rest of these notes. (Proofs can be found in [A1]. See also [F1].)

It follows from the fundamental theorem of algebra that every polynomial $p(z)$ with complex coefficients factors into linear factors:

$$p(z) = a_n z^n + a_{n-1} z^{n-1} + \cdots + a_1 z + a_0 = a_n (z - r_1)(z - r_2) \ldots (z - r_n),$$

where the roots $r_1, \ldots, r_n$ are in $\mathbb{C}$. The implication for polynomials with real coefficients is explained in the following corollary:

Corollary. *Any non-constant polynomial with real coefficients that does not factor into more than one factor is either linear or quadratic.*

This corollary is the key to proving the uniqueness of the complex numbers as stated in the following theorem:

3.3. Uniqueness of the Complex Numbers

Theorem 3.3.1. *Every field that is finite-dimensional over the reals is either isomorphic to $\mathbb{R}$ or to $\mathbb{C}$.*

Outline of the proof: Let S be a field that is a finite-dimensional vector space over the reals. The proof strategy is: (1) show that S contains the reals $\mathbb{R}$, (2) show that if S is larger than $\mathbb{R}$, then S contains the complex numbers $\mathbb{C}$, and (3) show that S is contained in $\mathbb{C}$.

Problems

1) Prove the corollary to the fundamental theorem of algebra stated above.

    (*Hint*: According to the fundamental theorem of algebra, a polynomial p can be factored into complex linear factors. The burden of the proof is to show that for a polynomial p, as described in the corollary, the number of linear factors is either one or two.

    First show that in the complex factorization of a polynomial with real coefficients the roots come in sets of conjugates; that is, if z is a root, then so is $\bar{z}$. Such a set of conjugates consists of either one real number or two complex numbers.

    Then verify that the product of the (one or two) linear factors arising from such a set of conjugate roots is a polynomial divisor of p with real coefficients (either linear or quadratic). Conclude that p must have only one set of conjugate roots, otherwise it would have a real factorization.)

2) Complete part (1) of the proof of Theorem 3.3.1.

    (*Hint*: Let 1 be the multiplicative identity of S. For every real x, let $i(x) = x1$. This multiplication is justified by the assumption that S is a vector space over the reals. Prove that i is an embedding of $\mathbb{R}$ into S. This proves that S contains an isomorphic copy of $\mathbb{R}$.)

3) Complete part (2) of the proof of Theorem 3.3.1.

    (*Hint*: Suppose that S is larger than the isomorphic copy of the reals that it contains. Let w be an element of S not in $\mathbb{R}$. Deduce that w is the root of a polynomial $p(x)$ with real coefficients by considering the sequence of elements $1, w, w^2, w^3$, etc, and using the assumption that S is finite-dimensional over $\mathbb{R}$.

    Use the corollary to the fundamental theorem of algebra, to conclude that there is a quadratic polynomial $p(x)$ with real coefficients and root

w. Show that $p(x) = x^2 + 2bx + c$ can be assumed to have leading coefficient 1.

Complete the square in the polynomial $p(x)$. Because w is not real, it follows that the discriminant $(b^2 - c)$ is negative, say $b^2 - c = -d^2$. (Why?). Conclude that $(w + b)/d$ is a square root of -1, so that S contains an isomorphic copy of the field $\mathbb{C}$.

4) Complete part (3) of the proof of Theorem 3.3.1.

(*Hint*: Let w be any element of S. Use the finite-dimensionality of S over the reals, to conclude that w is the root of a polynomial $p(x)$ with real coefficients. Apply the fundamental theorem of algebra to conclude that w is complex. This proves that S is contained in the field $\mathbb{C}$.

More two-dimensional number systems

Although the complexes are the only multi-dimensional field, there are other interesting two-dimensional number systems. In this section we define some and invite the reader to explore them.

Let $\mathbb{R}(\mathbf{j})$ be the set of numbers of the form $a + b\mathbf{j}$, where a and b are real numbers, and $\mathbf{j}$ is, for the moment, an undefined symbol. We first stipulate that addition in $\mathbb{R}(\mathbf{j})$ follows the natural rule:

$$(a + b\mathbf{j}) + (c + d\mathbf{j}) = (a + c) + (b + d)\mathbf{j},$$

which is almost implied by our use of the plus sign. For multiplication, we insist that the real numbers inside $\mathbb{R}(\mathbf{j})$ multiply as they usually do and that the distributive law continues to hold. Then we can calculate:

$$(a + b\mathbf{j})(c + d\mathbf{j}) = a(c + d\mathbf{j}) + b\mathbf{j}(c + d\mathbf{j}) = ac + ad\mathbf{j} + b\mathbf{j}c + b\mathbf{j}d\mathbf{j}.$$

Finally, supposing that $\mathbf{j}$ commutes with real numbers, we get:

$$(a + b\mathbf{j})(c + d\mathbf{j}) = ac + ad\mathbf{j} + bc\mathbf{j} + bd\mathbf{j}^2$$

To complete the definition of multiplication for $\mathbb{R}(\mathbf{j})$, we must assign a meaning to the symbol $\mathbf{j}^2$. At a minimum we want $\mathbb{R}(\mathbf{j})$ to be closed under multiplication, but this only means that $\mathbf{j}^2$ must be in $\mathbb{R}(\mathbf{j})$, say

$$\mathbf{j}^2 = p + q\mathbf{j},$$

3.3. Uniqueness of the Complex Numbers 69

where p and q are real. Then we have

$$(a + b\mathbf{j})(c + d\mathbf{j}) = ac + ad\mathbf{j} + bc\mathbf{j} + bd(p + q\mathbf{j})$$
$$= (ac + bdp) + (ad + bc + bdq)\mathbf{j}.$$

There are three natural choices for p and q, namely (a) $\mathbf{j}^2 = -1$, (b) $\mathbf{j}^2 = 1$, and (c) $\mathbf{j}^2 = 0$. In case (a), $\mathbb{R}(\mathbf{j})$ is the complex numbers; in (b), $\mathbb{R}(\mathbf{j})$ is called the **double numbers**; in case (c), $\mathbb{R}(\mathbf{j})$ is called the **dual numbers**.

Problems

1) Regardless of the choice of $\mathbf{j}^2$, verify that $\mathbb{R}(\mathbf{j})$ satisfies the commutative and associative laws of multiplication.

2) Regardless of the choice of $\mathbf{j}^2$, verify that $\mathbb{R}(\mathbf{j})$ has a multiplicative identity.

3) Which double numbers have multiplicative inverses? How do the double numbers fail to be a field?

4) Which dual numbers have multiplicative inverses? How do the dual numbers fail to be a field?

It might appear that $\mathbb{R}(\mathbf{j})$ provides an infinite number of algebraic systems, but the three examples—the complex numbers, the double numbers and the dual numbers—are the only truly distinct ones. To prove this, start with

$$\mathbf{j}^2 - q\mathbf{j} = p,$$

and complete the square, getting

$$\left(\mathbf{j} - \frac{q}{2}\right)^2 = p + \frac{q^2}{4}.$$

There are now three cases, depending whether the real $p + q^2/4$ is positive, negative, or zero. These correspond to the three systems: the double, complex, and dual numbers.

Problems

5) Explain why completing the square is justified in all the number systems $\mathbb{R}(\mathbf{j})$.

6) If $p + q^2/4$ is positive, show that $\mathbb{R}(\mathbf{j})$ is isomorphic to the double numbers.

7) If $p + q^2/4$ is negative, show that $\mathbb{R}(\mathbf{j})$ is isomorphic to the complex numbers.

8) If $p + q^2/4$ is zero, show that $\mathbb{R}(\mathbf{j})$ is isomorphic to the dual numbers.

3.4 Complex Calculus

Complex functions

A complex function $w = f(z)$ of a complex variable involves four real variables: two for $z = x + iy$ and two for $w = f(z) = u + iv$. To visualize $f(z)$ as a transformation, it is necessary to draw two planes, one for z the other for w.

Figure 3.4.1 illustrates this for $w = f(z) = 2z$. For example, the point $z = 2 + i$ with real coordinates $x = 2, y = 1$, is transformed to the point $w = 4 + 2i$ with real coordinates $u = 4, v = 2$. Judging from the picture, the function $w = 2z$ is a stretching (or magnification) of the plane.

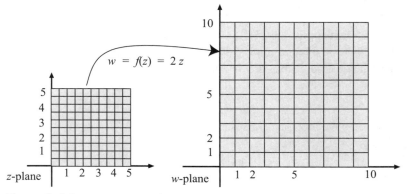

Figure 3.4.1. Graph of the function $w = 2z$ on the domain $0 \leq \text{Re}(z), \text{Im}(z) \leq 5$.

Problems

1) Graph these complex functions.

   (a) $f(z) = z + (2 + i)$

(b) $f(z) = z/2$
(c) $f(z) = iz$
(d) $f(z) = (i + z)/i$
(e) $f(z) = (1 + i)z$
(f) $f(z) = iz - i + 1$

2) Explain why the geometric interpretation given for each of these functions is correct.

   (a) $f(z) = z + a$, a real, is a horizontal translation of the plane.
   (b) $f(z) = z + bi$, b real, is a vertical translation of the plane.
   (c) $f(z) = z + z_0$, z_0 is a constant, is an arbitrary translation.
   (d) $f(z) = e^{i\theta}z$ is a rotation of the plane about the origin by the angle θ.

3) The function $f(z) = kz$ (where $k > 0$) is called a **homothetic** transformation. Describe its action on the complex plane.

4) What transformation does the function $f(z) = -z$ represent?

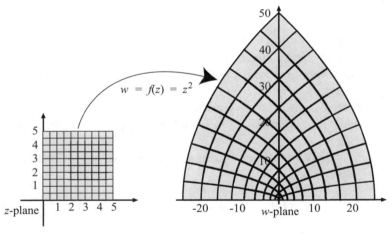

Figure 3.4.2. Graph of the function $w = z^2$.

5) The graph of $f(z) = z^2$ is given in Figure 3.4.2. Find equations for the curves shown in the w-plane. Verify that they are parabolas with focus at the origin.

   (*Hint*: In the complex equation
   $$u + iv = w = f(z) = (x + iy)^2,$$

pretend that x is a constant. Derive two real equations by separately equating the real and imaginary parts of this one. Eliminate y to get a single equation in u and v. Repeat, interchanging x and y.)

6) Graph $w = f(z) = 1/z$ on the square $0 \le x, y \le 5$, creating a figure like Figure 3.4.2. Find equations for the curves in the w-plane. Verify that they are circles passing through the origin.

Complex differentiation

The definition of the derivative for a complex function is the same as for a real function.

Definition. Let the function $f(z)$ be defined on a rectangle in the complex plane. Let z be a point therein. Then f is **differentiable** at z with derivative $f'(z)$ if the following limit exists:

$$f'(z) = \lim_{h \to 0} \frac{f(z+h) - f(z)}{h}.$$

Algebraically, the limit used here is the same as for real functions, so most of the same differentiation rules apply. In particular, the complex derivative obeys the sum, product, and chain rules. In other ways, complex analysis is very different from real analysis. A complex function with a derivative is much more special among complex functions of a complex variable than a differentiable function is among real-valued functions of a real variable.

To understand this, we separate a complex function $f(z)$ into real and imaginary parts:

$$f(z) = f(x + iy) = u(x, y) + iv(x, y),$$

where $u = \text{Re}(f)$ and $v = \text{Im}(f)$. Both u and v are *real*-valued functions of the *two real* variables x and y. Now the real limit as $h \to 0$ is taken two ways: the real number h approaches 0 from above and from below. If the limit exists, it is because both give the same result. Complex numbers are two-dimensional, so the complex limit as $h \to 0$ can be taken many ways. A complex h can approach 0 along any curve in the complex plane ending at 0. For the complex limit to exist, all ways of taking the limit must give the same result. This idea leads to the following theorem:

3.4. Complex Calculus

Theorem 3.4.1. *The real and imaginary parts, u and v, of a complex, differentiable function satisfy the* **Cauchy-Riemann Equations***:*

$$\frac{\partial u}{\partial x} = \frac{\partial v}{\partial y}, \quad \frac{\partial u}{\partial y} = -\frac{\partial v}{\partial x}$$

Part of the conclusion of this theorem is that the real and imaginary parts of a differentiable function of a complex variable are themselves differentiable real-valued functions (or at least have partial derivatives). By itself, this is not surprising. What is surprising is the link between the partial derivatives described by the Cauchy-Riemann equations.

Problems

7) Prove Theorem 3.4.1

(*Hint*: Let $h = x + iy$. One way that h can approach 0 is to let $h = x$, that is keep y fixed at 0. If we do this, then we get this formula for $f'(z)$:

$$f'(z) = \frac{\partial u}{\partial x}(z) + i\frac{\partial v}{\partial x}(z).$$

(Why?) On the other hand, we can let $h = iy$, keeping x at 0. Carefully work out a formula for $f'(z)$ in this second case and compare it with the previous result.)

8) Separate these complex functions into real and imaginary parts. Determine which are differentiable and check the Cauchy-Riemann equations.

(a) $f(z) = z^2$
(b) $f(z) = |z|$
(c) $f(z) = e^z$
(d) $f(z) = \overline{z+1}$
(e) $f(z) = 1/(z+1)$
(f) $f(z) = z^3$

9) Prove that the real and imaginary parts (u and v) of a differentiable function of a complex variable separately satisfy Laplace's equation:

$$\frac{\partial^2 w}{\partial x^2} + \frac{\partial^2 w}{\partial y^2} = 0.$$

Verify this for the functions in problem 8.

Conformality

The Cauchy-Riemann equations are the algebraic expression of the special nature of a complex differentiable function; conformality is the geometric expression. A **conformal** mapping is a transformation of the plane that preserves angles between curves. Figure 3.4.3 illustrates the difference between a conformal mapping and a non-conformal mapping when applied to a square grid. The lines of this grid in the z-plane meet at right angles. The conformal map preserves these angles, but the non-conformal map distorts them.

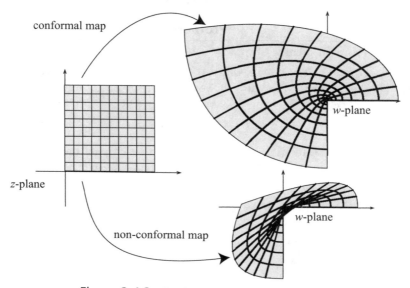

Figure 3.4.3. Conformal vs. non-conformal maps.

After seeing this illustration and comparing it with the other figures in this section, the reader will not be surprised by the next theorem:

Theorem 3.4.2. *A differentiable function of a complex variable is conformal.*

Proof. Let $z(t) = x(t) + iy(t)$ represent a planar curve γ parametrically and let Γ be the transformed curve $f(\gamma)$ (see Figure 3.4.4).

The derivative $z'(0) = x'(0) + iy'(0)$ gives the direction and magnitude of the tangent vector to the curve at $t = 0$, as in the figure. Applying the transformation $f(z)$ we get a curve $w(t) = f(z(t))$ in the w-plane.

3.4. Complex Calculus

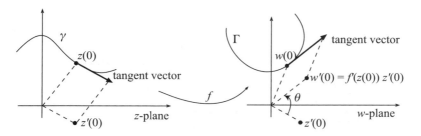

Figure 3.4.4. Proof of conformality.

According to the chain rule, $w'(0) = f'(z(0))z'(0)$. Therefore the tangent vector on the transformed curve is found algebraically by multiplying the original tangent vector ($z'(0)$) by the derivative of f at $z(0)$.

Let $f'(z(0)) = re^{i\theta}$. Multiplying this times $z'(0)$ stretches (or shrinks) $z'(0)$ by r and rotates it by θ, according to problem 5 in Section 3.2.

Suppose now that two parametric curves intersect in the z-plane at an angle φ. After transformation by $f(z)$, their tangents will both have been rotated by the same angle, so the transformed curves will continue to meet at the angle φ. (Draw your own graph.) This proves that f is conformal. □

Applications of Complex Functions

Figure 3.4.5 contains a graph of the complex square root $f(z) = \pm\sqrt{z}$. The lines parallel to the x-axis become hyperbolas with the coordinate axes as asymptotes; the lines parallel to the y-axis become hyperbolas with the 45° lines as asymptotes. The square root is conformal so these hyperbolas meet at right angles. Note the striking complexity of complex square roots compared with real square roots.

Figure 3.4.5 is also a typical application of complex functions. The w-plane depicts a **quadrupole field**, the electrostatic field produced by four plates carrying electrical charges, two positive and two negative, placed as indicated. The hyperbolas with the axes as asymptotes are the **field lines** or **lines of force**; the other hyperbolas are **equipotential lines**, curves that connect points with the same potential energy. A quadrupole field is, for example, used to focus electron beams. Calculations of the strength and shape of the quadrupole field use the complex square root. Fields created by plates with other shapes require other complex functions.

The graphs of complex functions describe many important families of curves including: **isotherms** (used to study heat flow), **isobars** (for the study of atmospheric pressure), **contour lines** (for the study of topography), and

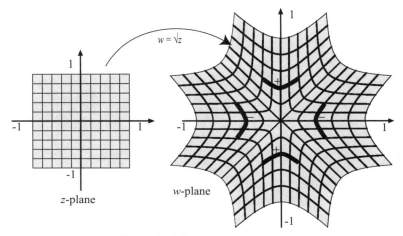

Figure 3.4.5. A quadrupole field.

stream lines (used to study fluid flow and the diffusion of gases). Complex functions have extensive practical applications in the design of objects that have to pass smoothly through a fluid. Airplane wings, automobile bodies, and submarine hulls are examples.

3.5 A Final Word about the Complexes

Although complex numbers were known and used from medieval times on, their geometric nature as points in a plane was not understood. Since their systematic treatment and application requires appreciation of their geometry, they languished for many years—as the name imaginary suggests—in a kind of mathematical limbo, occasionally used but not fully accepted.

It was Johann Carl Friederich Gauss (1777–1855) who first understood the geometry of the complex numbers and used them with confidence, giving in 1799 the first proof of the fundamental theorem of algebra. Complex numbers still required many more decades before they won complete acceptance. Their theory is now well-established and their use indispensible in many areas.

References for the complex numbers: This is the favorite subject of so many mathematicians that there is a host of excellent books among which [F1] is outstanding. [F3] is also excellent and includes a good discussion of applications; [F4] emphasizes the geometry of the complexes.

4

The Quaternions

4.1 Four-Dimensional Algebra and Geometry

Skew fields

It is disappointing that $\mathbb{C}$ is the only multi-dimensional field containing the reals. It would particularly be useful if some kind of multiplication could be defined in three-dimensional space making it a field. (We live, after all, in three-dimensional space.)

Unfortunately, this is impossible. However, if we give up one field axiom, we can proceed.

Definition. A **skew field** is a set S together with operations of addition and multiplication satisfying all the axioms of a field except the commutative law of multiplication.

There is no skew field of three-dimensions, sad to say, but there is a four-dimensional skew field. As with the complex numbers, it is convenient to start with matrices.

Definition. A **quaternion** is a 4 by 4 matrix of the form

$$\mathbf{q} = \begin{bmatrix} a & b & c & d \\ -b & a & -d & c \\ -c & d & a & -b \\ -d & -c & b & a \end{bmatrix},$$

where $a, b, c,$ and d are real numbers. Let $\mathbb{W}$ be the set of all quaternions.

4. The Quaternions

Like the complex numbers, $\mathbb{W}$ inherits operations of addition and multiplication from the algebra of matrices. These operations already satisfy most of the field axioms.

Theorem 4.1.1. $\mathbb{W}$ *is a skew field.*

If we introduce the matrices

$$\mathbf{i} = \begin{bmatrix} 0 & 1 & 0 & 0 \\ -1 & 0 & 0 & 0 \\ 0 & 0 & 0 & -1 \\ 0 & 0 & 1 & 0 \end{bmatrix}, \mathbf{j} = \begin{bmatrix} 0 & 0 & 1 & 0 \\ 0 & 0 & 0 & 1 \\ -1 & 0 & 0 & 0 \\ 0 & -1 & 0 & 0 \end{bmatrix}, \text{ and } \mathbf{k} = \begin{bmatrix} 0 & 0 & 0 & 1 \\ 0 & 0 & -1 & 0 \\ 0 & 1 & 0 & 0 \\ -1 & 0 & 0 & 0 \end{bmatrix}$$

then a quaternion can be written in the **Cartesian form** $a + b\mathbf{i} + c\mathbf{j} + d\mathbf{k}$, where the matrix

$$\begin{bmatrix} a & 0 & 0 & 0 \\ 0 & a & 0 & 0 \\ 0 & 0 & a & 0 \\ 0 & 0 & 0 & a \end{bmatrix}$$

is abbreviated by a. Looked at this way, a quaternion is like a complex number with three imaginary parts. The next definition makes this analogy even clearer.

Definition. Let $\mathbf{q} = a + x\mathbf{i} + y\mathbf{j} + z\mathbf{k}$ be a quaternion, where a, x, y, and z are real numbers. The **scalar part** of $\mathbf{q}$ is defined

$$S\mathbf{q} = a,$$

and the **vector part** of $\mathbf{q}$ is

$$V\mathbf{q} = x\mathbf{i} + y\mathbf{j} + z\mathbf{k}.$$

We also have the **conjugate**

$$\overline{\mathbf{q}} = S\mathbf{q} - V\mathbf{q} = a - x\mathbf{i} - y\mathbf{j} - z\mathbf{k},$$

and the **norm**

$$|\mathbf{q}| = (a^2 + x^2 + y^2 + z^2)^{1/2}.$$

If $S\mathbf{q} = 0$, then $\mathbf{q}$ is called a **pure** quaternion. If $|\mathbf{q}| = 1$, then $\mathbf{q}$ is a **unit** quaternion.

4.1. Four-Dimensional Algebra and Geometry

Problems

1) Show that the quaternions **i, j, k** satisfy:
   (a) $\mathbf{i}^2 = \mathbf{j}^2 = \mathbf{k}^2 = -1$
   (b) $\mathbf{ij} = -\mathbf{ji} = \mathbf{k}$
   (c) $\mathbf{jk} = -\mathbf{kj} = \mathbf{i}$
   (d) $\mathbf{ki} = -\mathbf{ik} = \mathbf{j}$

   (*Note*: Multiplication of quaternions is not commutative!)

2) Prove Theorem 4.1.1.

3) Find the Cartesian form of these quaternions.
   (a) **ijk**
   (b) $(2\mathbf{i} + \mathbf{j})(\mathbf{j} + \mathbf{k})$
   (c) $(2 + 3\mathbf{j})(\mathbf{i} + \mathbf{j} - \mathbf{k})$

4) Prove:
   (a) $\overline{\mathbf{qr}} = \overline{\mathbf{q}}\,\overline{\mathbf{r}}$
   (b) $|\mathbf{q}|^2 = \mathbf{q}\overline{\mathbf{q}}$
   (c) $|\mathbf{qr}| = |\mathbf{q}||\mathbf{r}|$
   (d) $|\mathbf{q} + \mathbf{r}| \leq |\mathbf{q}| + |\mathbf{r}|$
   (e) $S\mathbf{q} = \frac{\mathbf{q} + \overline{\mathbf{q}}}{2}$
   (f) $V\mathbf{q} = \frac{\mathbf{q} - \overline{\mathbf{q}}}{2}$
   (g) $\mathbf{q}^{-1} = \frac{\overline{\mathbf{q}}}{|\mathbf{q}|^2}$

5) Prove that $(\mathbf{qr})^{-1} = \mathbf{r}^{-1}\mathbf{q}^{-1}$.

6) Find the Cartesian form of
   (a) $(1 + \mathbf{i})^{-1}$
   (b) $(\mathbf{i} + \mathbf{j} + 2\mathbf{k})^{-1}$
   (c) $(2 - \mathbf{i} + \mathbf{j})^{-1}$

7) Show that the reciprocal of a pure quaternion is a pure quaternion.

8) Show that a **normalized** quaternion $\mathbf{q}/|\mathbf{q}|$ is a unit quaternion.

9) Is $\mathbb{W}$ Cauchy complete?

10) Do the quaternions satisfy the integral domain property?

11) Verify that all pure, unit quaternions are square roots of -1. Prove that these are the only square roots of -1.

12) There are many more than two quaternion square roots of -1 (see the previous exercise). What, then, is wrong with this argument? Let $\mathbf{q}$ be a solution of the equation $\mathbf{q}^2 = -1$. Rewriting we get $\mathbf{q}^2 + 1 = 0$. Factoring gives $(\mathbf{q} - \mathbf{i})(\mathbf{q} + \mathbf{i}) = 0$, therefore $\mathbf{q} = \mathbf{i}$ or $\mathbf{q} = -\mathbf{i}$ and we see that there are exactly two quaternion square roots of -1.

Dot and cross product

A pure quaternion $\mathbf{q} = x\mathbf{i} + y\mathbf{j} + z\mathbf{k}$ represents a point in three-dimensional space. Multivariable calculus gives us two multiplications for 3-dimensional vectors: the dot and the cross product. The next theorem interprets the multiplication of pure quaternions in these terms.

Theorem 4.1.2. *Let $\mathbf{q}$ and $\mathbf{r}$ be pure quaternions:*

$$\mathbf{q} = x_1\mathbf{i} + y_1\mathbf{j} + z_1\mathbf{k},$$

and

$$\mathbf{r} = x_2\mathbf{i} + y_2\mathbf{j} + z_2\mathbf{k}.$$

Then

$$\begin{aligned} S(\mathbf{qr}) &= x_1 x_2 + y_1 y_2 + z_1 z_2 \\ &= (\mathbf{q} \cdot \mathbf{r}) \\ &= |\mathbf{q}||\mathbf{r}| \cos(\phi), \qquad \text{— the \textbf{dot product}} \end{aligned}$$

and

$$\begin{aligned} V(\mathbf{qr}) &= (y_1 z_2 - y_2 z_1)\mathbf{i} - (x_1 z_2 - x_2 z_1)\mathbf{j} + (x_1 y_2 - x_2 y_1)\mathbf{k} \\ &= \mathbf{q} \times \mathbf{r} \\ &= |\mathbf{q}||\mathbf{r}| \sin(\phi)\mathbf{u}. \qquad \text{— the \textbf{cross product}} \end{aligned}$$

Here ϕ is the angle from the vector $\mathbf{q}$ to the vector $\mathbf{r}$ and the pure unit quaternion $\mathbf{u}$ represents a unit vector in three-dimensional space perpendicular to the plane of the vectors represented by $\mathbf{q}$ and $\mathbf{r}$ and forming a right-handed system with them.

4.1. Four-Dimensional Algebra and Geometry

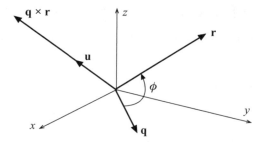

Figure 4.1.1. Direction **u** of the cross product **q** × **r** in relation to **q** and **r**.

In summary,

$$\mathbf{qr} = \overbrace{-(\mathbf{q}\cdot\mathbf{r})}^{\text{scalar part}} + \overbrace{(\mathbf{q}\times\mathbf{r})}^{\text{vector part}},$$

which gives a geometric interpretation for the multiplication of pure quaternions, because the dot and cross products themselves have simple geometric interpretations.

Problems

13) What form does the product of two pure quaternions take when they are perpendicular vectors in 3-dimensional space? When they are parallel vectors?

14) Verify, using quaternions:
    (a) $(\mathbf{q}\cdot\mathbf{r}) = (\mathbf{r}\cdot\mathbf{q})$,
    (b) $\mathbf{q}\times\mathbf{r} = -\mathbf{r}\times\mathbf{q}$,
    (c) $|\mathbf{q}\times\mathbf{r}|^2 = |\mathbf{q}|^2|\mathbf{r}|^2 - (\mathbf{q}\cdot\mathbf{r})^2$,

    where **q** and **r** are 3-dimensional vectors.

15) Use quaternion multiplication to find a formula for the area of the triangle determined by three points **q**, **r**, and **s**.

    (*Hint*: What is $S(\mathbf{qrs})$?)

4.2 The Polar Form of a Quaternion

A quaternion has a polar form analogous to the polar form of a complex number.

Theorem 4.2.1. *Every quaternion can be represented in the form*

$$\mathbf{q} = |\mathbf{q}|(\cos(\theta) + \mathbf{u}\sin(\theta)),$$

where $\mathbf{u}$ *is a pure, unit quaternion, and* $0 \leq \theta < 2\pi$.

Problems

1) Prove Theorem 4.2.1.

   (*Hint*: $\mathbf{u}$ is the normalized vector part of $\mathbf{q}$.)

2) The polar form of a quaternion features a pure, unit quaternion. What complex number or numbers plays the role of the pure, unit quaternion in the polar form of a complex number?

3) Is the polar form of a quaternion unique?

Geometric application of quaternions

Many geometric transformations of 3- and 4-dimensional space can be written compactly using the algebra of quaternions. Polar form plays a crucial role. Here is an example.

Theorem 4.2.2. *Let* $\mathbf{r} = \cos(\theta) + \mathbf{u}\sin(\theta)$, *where* $\mathbf{u}$ *is a pure unit quaternion (i.e., a unit vector in 3-dimensional space). Define algebraically the operation R on a vector* $\mathbf{q}$ *by*

$$R(\mathbf{q}) = \mathbf{rqr}^{-1}.$$

Then geometrically $R(\mathbf{q})$ *is the rotation of* $\mathbf{q}$ *around the axis* $\mathbf{u}$ *by the angle* 2θ. *Every rotation of 3-dimensional space about an axis passing through the origin can be represented in this way.*

Proof. Let T be the geometric transformation of three-dimensional space consisting of a rotation of 2θ about the axis $\mathbf{u}$. The goal is to show that $T = R$.

4.2. The Polar Form of a Quaternion

We first prove that T and R are linear transformations; that is, that

$$T(\mathbf{q}+\mathbf{s}) = T(\mathbf{q}) + T(\mathbf{s}), \qquad R(\mathbf{q}+\mathbf{s}) = R(\mathbf{q}) + R(\mathbf{s}),$$

and

$$T(\alpha\mathbf{q}) = \alpha T(\mathbf{q}), \qquad R(\alpha\mathbf{q}) = \alpha R(\mathbf{q}),$$

where $\mathbf{q}$ and $\mathbf{s}$ are arbitrary pure quaternions and α is a scalar (a real number).

For T, these properties are geometrically obvious. For example, if $\mathbf{q}$ is a quaternion and α is a scalar, it makes no difference whether we multiply $\mathbf{q}$ by α first (which stretches or compresses $\mathbf{q}$) and then rotate about $\mathbf{u}$, or rotate first and then multiply. For R, linearity follows by algebraic computation:

$$R(\mathbf{q}+\mathbf{s}) = \mathbf{r}(\mathbf{q}+\mathbf{s})\mathbf{r}^{-1} = \mathbf{rqr}^{-1} + \mathbf{rsr}^{-1} = R(\mathbf{q}) + R(\mathbf{s}),$$

and

$$R(\alpha\mathbf{q}) = \mathbf{r}\alpha\mathbf{q}\mathbf{r}^{-1} = \alpha\mathbf{rqr}^{-1} = \alpha R(\mathbf{q}),$$

using the distributive law for the first equation and the fact that real numbers commute with quaternions for the second.

To apply linearity, we choose a basis consisting of the vector $\mathbf{u}$, plus two orthogonal vectors, and proceed to check that $R = T$ for these basis vectors. It then follows that $R = T$ for all vectors.

(a) If $\mathbf{q} = \mathbf{u}$, then

$$R(\mathbf{u}) = \mathbf{rur}^{-1} = (\cos(\theta) + \mathbf{u}\sin(\theta))\mathbf{u}(\cos(\theta) - \mathbf{u}\sin(\theta))$$
$$= \cos(\theta)^2\mathbf{u} - \sin(\theta)^2\mathbf{u}^3 = \cos(\theta)^2\mathbf{u} - \sin(\theta)^2(-\mathbf{u}) = \mathbf{u}.$$

Likewise $T(\mathbf{u}) = \mathbf{u}$, since the axis of rotation is fixed in place by rotation.

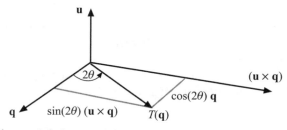

Figure 4.2.1. Transformation of a vector $\mathbf{q}$ perpendicular to $\mathbf{u}$.

(b) If **q** is in the plane perpendicular to **u** (see Figure 4.2.1), then

$$R(\mathbf{q}) = \mathbf{rqr}^{-1} = (\cos(\theta) + \mathbf{u}\sin(\theta))\mathbf{q}(\cos(\theta) - \mathbf{u}\sin(\theta))$$
$$= \cos(\theta)^2 \ \mathbf{q} + \mathbf{uq}\cos(\theta)\sin(\theta) - \mathbf{qu}\cos(\theta)\sin(\theta)$$
$$- \mathbf{uqu}\sin(\theta)^2.$$

Because **u** and **q** are perpendicular,

$$\mathbf{uq} = -(\mathbf{u} \cdot \mathbf{q}) + (\mathbf{u} \times \mathbf{q}) = (\mathbf{u} \times \mathbf{q}),$$

and

$$\mathbf{qu} = (\mathbf{q} \times \mathbf{u}) = -(\mathbf{u} \times \mathbf{q}),$$

so that

$$\mathbf{uqu} = (\mathbf{u} \cdot (\mathbf{u} \times \mathbf{q})) - (\mathbf{u} \times (\mathbf{u} \times \mathbf{q})) = -(\mathbf{u} \times (\mathbf{u} \times \mathbf{q})),$$

since $(\mathbf{u} \times \mathbf{q})$ is also perpendicular to **u**. Using the right-hand rule, we see that

$$\mathbf{uqu} = \mathbf{q}.$$

Putting all this together, gives

$$R(\mathbf{q}) = \cos(\theta)^2 \mathbf{q} + 2\cos(\theta)\sin(\theta)(\mathbf{u} \times \mathbf{q}) - \sin(\theta)^2 \mathbf{q}$$
$$= \cos(2\theta)\mathbf{q} + \sin(2\theta)(\mathbf{u} \times \mathbf{q}),$$

therefore $R(\mathbf{q})$ is the rotation of **q** through an angle of 2θ about the axis **u**, that is,

$$R(\mathbf{q}) = T(\mathbf{q}).$$

□

Problems

4) Prove that two three-dimensional vectors **q** and **r** represented by pure quaternions are orthogonal if and only if

$$\mathbf{qr} = -\mathbf{rq},$$

that is, anti-commuting vectors are orthogonal.

4.2. The Polar Form of a Quaternion

5) Let **r** be any quaternion. Prove that the transformation

$$R(\mathbf{q}) = \mathbf{rqr}^{-1}$$

of the pure quaternion **q** is a rotation in 3-dimensional space.

6) Let $\mathbf{r}_1$ and $\mathbf{r}_2$ be quaternions. Let R_1 and R_2 be the corresponding rotations *a la* Theorem 4.2.2. Show that R_1 composed with R_2 corresponds to the quaternion $\mathbf{r}_1 \mathbf{r}_2$.

7) Prove that the composition of two rotations of 3-dimensional space is another rotation.

8) Let R_1 be a rotation of 90 degrees about an axis $\mathbf{u}_1$. Let R_2 be a second rotation of 90 degrees about an axis $\mathbf{u}_2$ perpendicular to $\mathbf{u}_1$. Using quaternions, find a formula for the composition R_1 and R_2. Describe the resulting transformation geometrically.

9) The preceding exercise points out that the rules governing the combination of rotations in three dimensions are complicated. In contrast, what is the rule governing the combination of rotations in two dimensions?

10) Find a formula for rotation of three-dimensional space about an axis not passing through the origin.

11) Let **r** be a pure unit quaternion. Let ρ be the plane in 3-dimensional space perpendicular to **r** and passing through the origin. Prove that

$$T(\mathbf{q}) = \mathbf{rqr}$$

is a reflection across the plane ρ.

(*Hint*: Imitate the proof of Theorem 4.2.2.)

Uniqueness of the quaternions

One might hope to find skew fields of still higher dimension, but there are no more, at least that are finite dimensional. (For a proof, see [A1] or [C2].)

Uniqueness of the Quaternions. *(Frobenius) Every skew field finite-dimensional over the reals is isomorphic to* $\mathbb{R}$ *or to* $\mathbb{C}$ *or to* $\mathbb{W}$.

4.3 Complex Quaternions and the Quaternion Calculus

Complex quaternions

Although there are no skew fields of dimension higher than four, there are some other interesting number systems. The complex quaternions are an example. They provide a model for the theory of relativity.

Definition. A **complex quaternion** is a quantity of the form

$$\mathbf{q} = t + i x \mathbf{i} + i y \mathbf{j} + i z \mathbf{k},$$

where i is the complex unit.

In other words $\mathbf{q}$ is a quaternion with one real and three imaginary coefficients (weird). As with ordinary quaternions, the **scalar part** of $\mathbf{q}$ is defined by

$$S\mathbf{q} = t.$$

the **vector part** by

$$V\mathbf{q} = ix\mathbf{i} + iy\mathbf{j} + iz\mathbf{k},$$

the **conjugate** of $\mathbf{q}$ by

$$\mathbf{q}^* = S\mathbf{q} - V\mathbf{q} = t - ix\mathbf{i} - iy\mathbf{j} - iz\mathbf{k},$$

and the (**Minkowski**) **norm** of $\mathbf{q}$ by

$$||\mathbf{q}|| = |t^2 - x^2 - y^2 - z^2|^{1/2}.$$

Absolute values are needed for the norm since $t^2 - x^2 - y^2 - z^2$ can be negative.

Problems

1) Prove that complex quaternions obey the associative and distributive laws.

2) Which of the axioms of a skew field don't hold for the complex quaternions?

3) Prove:

   (a) $(\mathbf{qr})^* = \mathbf{r}^*\mathbf{q}^*,$

4.3. Complex Quaternions and the Quaternion Calculus

(b) $||\mathbf{q}||^2 = |\mathbf{q}\mathbf{q}^*| = |\mathbf{q}^*\mathbf{q}|$,

(c) $||\mathbf{q}\mathbf{r}|| = ||\mathbf{r}\mathbf{q}||$.

4) Find a nonzero complex quaternion with $||\mathbf{q}|| = 0$.

5) Verify that not every nonzero complex quaternion has an inverse. Show that complex quaternions do not satisfy the cancellation law. Do the quaternions satisfy the integral domain property?

6) Prove that a complex quaternion of nonzero Minkowski norm can be written in the form

$$\mathbf{q} = ||\mathbf{q}||(\cosh(\tau) + i\mathbf{u}\sinh(\tau)),$$

where $\mathbf{u}$ is a pure unit quaternion. This is **polar form** for the complex quaternions.

(*Hint*: The functions cosh and sinh parametrize the unit hyperbola $x^2 - y^2 = 1$, in the same way that cos and sin parametrize the unit circle. If $x^2 - y^2 = 1$, and $x > 0$, there is a τ such that $(\cosh(\tau), \sinh(\tau)) = (x, y)$.)

Lorentz transformations

The transformation $L(\mathbf{w}) = \mathbf{q}\mathbf{w}\mathbf{q}$ where $||\mathbf{q}|| = 1$ is called a **Lorentz transformation**. Lorentz transformations are a kind of rotation in the 4-dimensional space of complex quaternions. Let us compute one. Thus let $\mathbf{q}$ be the particular unit quaternion

$$\mathbf{q} = (\cosh(\tau) + i\mathbf{j}\sinh(\tau)),$$

where, for simplicity, $\mathbf{q}$ has been given only two nonzero components. If $\mathbf{w}$ is a typical complex quaternion

$$\mathbf{w} = t + ix\mathbf{i} + iy\mathbf{j} + iz\mathbf{k},$$

then

$$L(\mathbf{w}) = \mathbf{q}\mathbf{w}\mathbf{q} = (\cosh(\tau) + i\mathbf{j}\sinh(\tau))(t + ix\mathbf{i} + iy\mathbf{j} + iz\mathbf{k}) \\ (\cosh(\tau) + i\mathbf{j}\sinh(\tau)).$$

4. The Quaternions

Multiplying out,

$$L(\mathbf{w}) = [(\cosh(\tau)^2 + \sinh(\tau)^2)t + 2\cosh(\tau)\sinh(\tau)y]$$
$$+ i(\cosh(\tau)^2 - \sinh(\tau)^2)x\mathbf{i}$$
$$+ i[2\cosh(\tau)\sinh(\tau)t + (\cosh(\tau)^2 + \sinh(\tau)^2)y]\mathbf{j}$$
$$+ i(\cosh(\tau)^2 - \sinh(\tau)^2)z\mathbf{k},$$

and using the hyperbolic identities:

$$\cosh(\tau)^2 - \sinh(\tau)^2 = 1,$$
$$\cosh(2\tau) = \cosh(\tau)^2 + \sinh(\tau)^2,$$
$$\sinh(2\tau) = 2\cosh(\tau)\sinh(\tau),$$

$L(\mathbf{w})$ becomes

$$L(\mathbf{w}) = [\cosh(2\tau)t + \sinh(2\tau)y] + ix\mathbf{i} + i[\sinh(2\tau)t + \cosh(2\tau)y]\mathbf{j} + iz\mathbf{k}.$$

Note the similarity between this formula and the corresponding formula for a Euclidean rotation (using ordinary quaternions):

$$R(\mathbf{v}) = \mathbf{qvq}^*$$
$$= (\cos(\theta) + \mathbf{j}\sin(\theta))(x\mathbf{i} + y\mathbf{j} + z\mathbf{k})(\cos(\theta) - \mathbf{j}\sin(\theta))$$
$$= [(\cos(\theta)^2 - \sin(\theta)^2)x + 2\cos(\theta)\sin(\theta)z]\mathbf{i}$$
$$+ (\cos(\theta)^2 + \sin(\theta)^2)y\mathbf{j}$$
$$+ [-2\cos(\theta)\sin(\theta)x + (\cos(\theta)^2 - \sin(\theta)^2)z]\mathbf{k}$$
$$= [\cos(2\theta)x + \sin(2\theta)z]\mathbf{i} + y\mathbf{j} + [-\sin(2\theta)x + \cos(2\theta)z]\mathbf{k},$$

where $\mathbf{v} = x\mathbf{i} + y\mathbf{j} + z\mathbf{k}$ is an ordinary pure quaternion and $\mathbf{q} = \cos(\theta) + \mathbf{j}\sin(\theta)$ is an ordinary unit quaternion.

It is the analogy between a Euclidean rotation and a Lorentz transformation suggests that the Lorentz transformation is a kind of hyperbolic rotation.

Problems

7) Check the algebraic details of the formulas for $L(\mathbf{w})$ and $R(\mathbf{v})$ given above.

8) Derive the hyperbolic identities used above.

4.3. Complex Quaternions and the Quaternion Calculus

The special theory of relativity

Einstein's theory of relativity describes physical events as seen by an ideal observer occupying what is called an inertial reference frame, essentially a position free of gravitational influence. Einstein maintained that under these special conditions (whence the term "special" relativity) the laws of physics would appear the same to all observers.

According to the special theory, physical events relative to an inertial frame of reference are described by four coordinates: time t and the three space coordinates x, y, and z. The resulting four-dimensional coordinate system is called **space-time**. The coordinates of this system correspond to the four components of a complex quaternion. Thus let

$$\mathbf{w}_1 = t_1 + ix_1\mathbf{i} + iy_1\mathbf{j} + iz_1\mathbf{k},$$

and

$$\mathbf{w}_2 = t_2 + ix_2\mathbf{i} + iy_2\mathbf{j} + iz_2\mathbf{k}$$

represent two events seen by an observer in an inertial frame of reference.

In classical physics there are two ways to measure the distance between events: time separation and spatial distance. In quaternion terms these are

$$\text{time separation} = |S(\mathbf{w}_1) - S(\mathbf{w}_2)| = |t_1 - t_2| = ((t_1 - t_2)^2)^{1/2},$$

and

$$\begin{aligned}\text{space separation} &= |V(\mathbf{w}_1) - V(\mathbf{w}_2)| \\ &= ((x_1 - x_2)^2 + (y_1 - y_2)^2 + (z_1 - z_2)^2)^{1/2}.\end{aligned}$$

In classical physics two observers of the events $\mathbf{w}_1$ and $\mathbf{w}_2$ would measure the same time separation and space separation.

In contrast, according to Einstein's theory, two observers of $\mathbf{w}_1$ and $\mathbf{w}_2$ do not necessarily observe the same time and space separations; they *will* observe the same **relativistic separation** or **Minkowski separation**:

$$||\mathbf{w}_1 - \mathbf{w}_2|| = |(t_1 - t_2)^2 - (x_1 - x_2)^2 - (y_1 - y_2)^2 - (z_1 - z_2)^2|^{1/2}.$$

Thus Minkowski separation is the appropriate distance function for the geometry of relativistic space-time.

The quantity

$$M = (t_1 - t_2)^2 - (x_1 - x_2)^2 - (y_1 - y - 2)^2 - (z_1 - z_2)^2$$

can be negative as well as positive or zero. This gives rise to three different types of intervals between events. If M is positive, then the time separation predominates and the interval between $\mathbf{w}_1$ and $\mathbf{w}_2$ is called **timelike**. If M is negative, then space separation predominates and the interval is called **spacelike**. Figure 4.3.1 illustrates these possibilities with the z dimension suppressed.

If M is zero, the interval is called **lightlike**. Lightlike intervals are observed when it is possible for a ray of light to travel between the two events. Units have been chosen so that the speed of light is 1. Any other particle travels at a rate of speed less than that of light (less than 1). Therefore in the interval separating two events along the path followed by non-light particles, time separation predominates over space separation. In other words, particles, other than rays of light, travel so as to create timelike intervals between events along their path through space-time.

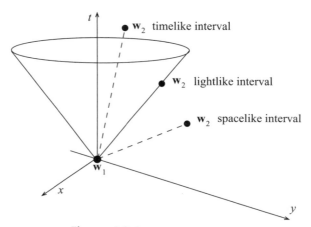

Figure 4.3.1. Relativistic intervals.

For two observers in inertial frames of reference, suppose that the reference frame of observer 2 is traveling at a constant speed of σ in relation to the frame of observer 1. For simplicity, suppose that the motion of observer 2 in relation to observer 1 is in the y direction. Based on physical arguments the relativistic transformation connecting the views $\mathbf{w}_1$ and $\mathbf{w}_2$ of the two observers is

4.3. Complex Quaternions and the Quaternion Calculus

$$\begin{cases} t_1 = \frac{t_2}{\sqrt{1-\sigma^2}} + \frac{y_2\sigma}{\sqrt{1-\sigma^2}}, \\ x_1 = x_2, \\ y_1 = \frac{t_2\sigma}{\sqrt{1-\sigma^2}} + \frac{y_2}{\sqrt{1-\sigma^2}}, \\ z_1 = z_2. \end{cases}$$

The important thing is to notice that the coefficients a and b of t_2 and y_2:

$$a = 1/(1-\sigma^2)^{1/2},$$
$$b = \sigma/(1-\sigma^2)^{1/2}$$

satisfy $a^2 - b^2 = 1$, therefore there is a number τ, called the **velocity parameter**, such that $a = \cosh(\tau)$ and $b = \sinh(\tau)$. Written using τ, it is apparent that the relativistic transformation is the same as the Lorentz transformation:

$$\mathbf{w}_1 = L(\mathbf{w}_2) = \mathbf{q}\mathbf{w}_2\mathbf{q},$$

where $\mathbf{q} = \cosh(\tau) + i \sinh(\tau)\mathbf{j}$. Thus a Lorentz transformation connects the viewpoints of two observers in relativistic space-time.

The calculus of quaternions

The best feature of the quaternion calculus is the **nabla** operator, invented by William Hamilton. It is defined as follows:

$$\nabla = \frac{\partial}{\partial x}\mathbf{i} + \frac{\partial}{\partial y}\mathbf{j} + \frac{\partial}{\partial z}\mathbf{k}.$$

The next theorem shows how this operator is connected with all the principal operators of vector calculus:

Theorem 4.3.1. *Let $f(x, y, z)$ be a real-valued function. Then ∇f is the* **gradient** *of f:*

$$\nabla f = \frac{\partial f}{\partial x}\mathbf{i} + \frac{\partial f}{\partial y}\mathbf{j} + \frac{\partial f}{\partial z}\mathbf{k},$$

and

$$\nabla^2 f = \frac{\partial^2 f}{\partial x^2} + \frac{\partial^2 f}{\partial y^2} + \frac{\partial^2 f}{\partial z^2}$$

is the **Laplacian**.

If
$$F(x, y, z) = u(x, y, z)\mathbf{i} + v(x, y, z)\mathbf{j} + w(x, y, z)\mathbf{k}$$
is a vector function, then
$$\nabla F = \text{curl}(F) + \text{div}(F),$$
where
$$\text{curl}(F) = \nabla \times F = \left(\frac{\partial w}{\partial y} - \frac{\partial v}{\partial z}\right)\mathbf{i} + \left(\frac{\partial u}{\partial z} - \frac{\partial w}{\partial x}\right)\mathbf{j} + \left(\frac{\partial v}{\partial x} - \frac{\partial u}{\partial y}\right)\mathbf{k},$$
and
$$\text{div}(F) = \nabla \cdot F = \frac{\partial u}{\partial x} + \frac{\partial v}{\partial y} + \frac{\partial w}{\partial z}.$$

The nabla operator applies only to real functions of real variables. What about functions of a quaternion variable? One way to define them is through power series. For example, here is a quaternion version of the exponential function.

Definition. The quaternion exponential function is defined by
$$\exp(\mathbf{q}) = 1 + \mathbf{q} + \frac{\mathbf{q}^2}{2} + \frac{\mathbf{q}^3}{3!} + \cdots + \frac{\mathbf{q}^n}{n!} + \cdots$$

Unfortunately, this function does not have the nice properties that the real and complex exponential functions have (see problem 11 below). Its difficulties stem from the fact that the quaternions are not commutative.

Turning to differentiation, a problem immediately arises with the derivative
$$\mathbf{f}'(\mathbf{q}) = \lim_{h \to 0} \frac{\mathbf{f}(\mathbf{q} + \mathbf{h}) - \mathbf{f}(\mathbf{q})}{\mathbf{h}},$$
where $\mathbf{f}(\mathbf{q})$ is quaternion function of a quaternion variable and the limit is taken as the quaternion $\mathbf{h}$ approaches the zero quaternion $\mathbf{0}$. The problem is rooted in the concept of a quaternion fraction. The usual notation for fractions p/q isn't appropriate because it does not discriminate between pq^{-1} and $q^{-1}p$, which may differ because quaternions don't necessarily commute. To form a proper difference quotient, we must choose whether to multiply $\mathbf{f}(\mathbf{q} + \mathbf{h}) - \mathbf{f}(\mathbf{q})$ by $\mathbf{h}^{-1}$ on the right or on the left. In other words, there are *two* quaternion derivatives.

Definition. Let $\mathbf{f}(\mathbf{q})$ be a quaternion function of a quaternion variable. The **right derivative** of $\mathbf{f}$ is given by
$$\mathbf{f}'_R(\mathbf{q}) = \lim_{h \to 0} (\mathbf{f}(\mathbf{q} + \mathbf{h}) - \mathbf{f}(a))\mathbf{h}^{-1},$$

4.3. Complex Quaternions and the Quaternion Calculus

while the **left derivative** is given by

$$f'_L(\mathbf{q}) = \lim_{\mathbf{h} \to 0} \mathbf{h}^{-1}(f(\mathbf{q} + \mathbf{h}) - f(\mathbf{q})).$$

The theory of these derivatives is quite unsatisfactory, as the problems show.

Problems

9) Prove Theorem 4.3.1.

10) Show that the infinite series for the quaternion exponential converges for any quaternion **q**.

11) When is the fundamental exponential law

$$\exp(\mathbf{q} + \mathbf{r}) = \exp(\mathbf{q}) \cdot \exp(\mathbf{r})$$

valid if **q** and **r** are quaternions?

12) Use the polar form of **q** to obtain another formula for exp(**q**). Find a connection between the quaternion and complex exponentials.

13) Using the left derivative alone, what differentiation rules can you prove? What about the product rule?

14) Find the left and right quaternion derivatives of these functions:
   (a) $f(\mathbf{q}) = \mathbf{jq} + \mathbf{k}$
   (b) $\exp(\mathbf{q})$

15) Show that the limit

$$\lim_{\mathbf{h} \to 0} (\mathbf{hqh}^{-1})$$

does not exist. Conclude that the simple function $f(\mathbf{q}) = \mathbf{q}^2$ has neither a left nor a right derivative.

(*Hint*: Let **q** be a simple pure quaternion and let **h** approach **0** first along the **i** axis and then along the **j** axis.)

4.4 A Final Word about the Quaternions

During the course of at least fifteen years spent thinking about the problem of finding a set of multi-dimensional numbers that would reflect the laws governing rotations in three dimensions, William Rowan Hamilton (1805–1865) came to realize that such a set of numbers would require four components not three and that the commutative law of multiplication, which had hitherto held for all systems of numbers, would have to be sacrificed. One evening in October 1843, the final element of the solution of the problem came to him while he walked along the Royal Canal in Dublin. This was the formula

$$\mathbf{i}^2 = \mathbf{j}^2 = \mathbf{k}^2 = \mathbf{ijk} = -1,$$

which determines the strange multiplication of the quaternions. With a pocket knife, Hamilton engraved this formula on the stone of Brougham Bridge (now called Broom Bridge). The bridge is still there and Hamilton's discovery is marked by a plaque.

As the first non-commutative and first non-planar set of numbers to be developed, the quaternions were a powerful influence on the development of algebra, contributing to the invention of matrix algebra later in the nineteenth century and the development of abstract algebra in the twentieth. The quaternions seem also to have anticipated some of the four-dimensional geometric ideas of the theory of relativity.

The quaternions were used for some years in the study of motion of three-dimensional objects. Great hopes were expressed for them, but no calculus of quaternions was forth-coming, and soon quaternions were supplanted by more general matrix and linear algebra methods. However, with the use of computers, quaternions have come to be a standard tool for executing three-dimensional rotations efficiently for the purpose of animation. Quaternions are also still in use in the investigation of fundamental geometric and topological problems in three and four dimensions.

References for the quaternions: [A1] and [C3] are good introductory sources. For more on quaternions and relativity, see [G4]. Finally, for the application of quaternions to animation, see [G8].

What number systems lurk beyond four dimensions? An important family of such systems are the geometric algebras (also called Clifford algebras). See [G5] and [G7] for these. The latter reference is an extensive bibliography.

Part III

ALTERNATIVE LINES

We turn to alternative real lines, number systems that contain the real numbers and have the same algebraic and geometric properties as the reals. If these systems are so similar, same algebra, same geometry, and even, to a certain extent, the same numbers, then the reader may well ask, why bother with them?

These systems are interesting not so much because they contain different numbers. They embody different ideas of number, radically different philosophies of mathematics. They have conflicting visions about what should be allowed to be a number, what properties are possible for numbers, and what tools should be permitted to prove those properties.

The three systems we present portray a kind of political spectrum of mathematical philosophies, from the radical right (the constructive reals), through moderately liberal (the hyperreals), to the radical left (the surreals). Each embodies a distinctive vision of what numbers are, how to calculate with them, and how to prove theorems about them.

5

The Constructive Reals

> God made the integers, everything else is the work of man.
> —Leopold Kronecker

> Building on the positive integers, weaving a web of ever more sets and more functions, we get the basic structures of mathematics: the rational number system, the real number system, the Euclidean spaces, the complex number system, the algebraic number fields, Hilbert space, the classical groups, and so forth. Within the framework of these structures most mathematics is done. Everything attaches itself to number, and every mathematical statement ultimately expresses the fact that if we perform certain computations within the set of positive integers, we shall get certain results. —Errett Bishop

5.1 Constructivist Criticism of Classical Mathematics

The constructive reals are the product of a radically conservative approach to mathematics. The constructivists take the integers as intuitively given, god given as Kronecker said, and the one and only source of truth in mathematics. To preserve this truth, they insist that all mathematical statements should be verifiable by computations within integers. The key idea here is that of a **computation**, by which is meant an operation or sequence of operations that can be performed by a finite intelligence (you or me, for example, or a digital computer) in a finite number of steps. By insisting that mathematics be computationally verifiable or reducible to finite computations with integers, the constructivists aim to guarantee truth in mathematics.

As an example of a mathematical statement that can be computationally verified we have the rational inequality

$$\frac{1}{2} < \frac{5}{7},$$

which can be verified by reducing it by cross multiplication to an integer inequality

$$7 < 10$$

that can be directly verified within the integers.

For a more complicated example, take the inequality

$$\sqrt{2} > 0.$$

In Cantor's construction of the reals, $\sqrt{2}$ is represented by a Cauchy sequence of rationals, $\{a_n\}$, for example, the sequence

$$\left\{2, \frac{3}{2}, \frac{17}{12}, \frac{577}{408}, \ldots\right\},$$

obtained inductively, using Newton's Method, by setting

$$a_1 = 2,$$

and, for $n > 1$,

$$a_n = \frac{1}{2}\left(a_{n-1} + \frac{2}{a_{n-1}}\right).$$

The problem of dealing with $\sqrt{2}$ is thus reduced to dealing with a particular sequence of rationals. The definition of positivity in Chapter 1 requires that we now find two integers M and N such that if $n > N$ then $a_n > 1/M$. In this case, we can easily prove by mathematical induction that $1 < a_n \leq 2$ for all $n \geq 1$, so that both M and N can be taken to be 1. This argument uses only operations on rational numbers, which can be reduced to operations on integers. Thus we have verified that

$$\sqrt{2} > 0.$$

by computations within the integers.

Problems

1) Using mathematical induction, prove that $1 < a_n \leq 2$, where $\{a_n\}$ is the sequence defined above.

2) Prove that $a_n^2 > 2$ for all n.

   (*Hint*: Consider $a_n^2 - 2$.)

3) Prove that the sequence $\{a_n\}$ converges.

   (*Hint*: Use mathematical induction to prove that $\{a_n\}$ is bounded and monotonically decreasing.)

4) Prove that $\{a_n\}$ converges to $\sqrt{2}$. Conclude that $1 < \sqrt{2} \leq 2$.

Constructive criticism

In the constructive view, some well-known mathematical ideas and theorems are meaningless or, worse, actually false. How is it that falsehood has crept into mathematics?

One source is the increasing abstraction of modern mathematics. During the last century mathematics has grown rapidly. By some estimates more than 90% of all known mathematics is the product of the twentieth century. Much recent mathematics is abstract in the sense that it refers to mathematical objects whose existence is hypothetical. For example, in linear algebra when a theorem begins

"Let V be a vector space ...,"

the existence of V is hypothetical. While in many cases there is no harm in this, critics have pointed out that the trend toward abstract mathematics tends to produce mathematics that is without computational basis; that is, the statements made cannot be confirmed by performing, in Errett Bishop's words, "computations within the set of positive integers." Such mathematics strays from the principle (not embraced by all mathematicians) that mathematical results should refer to and be verified by computation.

Thus, for example,

"Let V be a vector space, then V has a basis."

is a mathematical statement that cannot be verified by computation because there is no way to construct (i.e., compute) a basis for many infinite-dimensional vector spaces. All proofs of this result for infinite dimensional vector spaces are non-constructive (see problems 5–7). This is one way abstract mathematics leads away from results that are strictly computable.

Problems

5) Let $\Re^\infty$ be the vector space of all infinite sequences of reals (that is, infinite vectors or infinite n-tuples of reals). Try to find a basis for $\Re^\infty$. Recall that a **basis** for a vector space is a set of vectors such that every vector in the space can be expressed as a *finite* linear combination of the basis vectors.

6) Try to find a specific basis for the vector space $\mathbb{C}[0, 1]$ consisting of all continuous real-valued functions $f(x)$ defined for $0 \leq x \leq 1$.

   (*Discussion*: No specific basis is known. Try to find one anyway, that is, consider how one might go about selecting basis functions.)

7) Try to prove that every vector space has a basis. With what can you begin? How can you proceed?

More constructive criticism

Abstract mathematics is not the only source of results that are not computationally verifiable. Take the completeness of the real number system:

> "Every bounded non-empty set of reals has a least upper bound."

Following Bishop [H1], consider the simpler statement:

> "Every bounded sequence $\{a_n\}$ of rational numbers has a real least upper bound b."

If this is constructively verifiable, then it should be possible, given a sequence $\{a_n\}$, to compute its least upper bound b, or at least rational approximations to b of any desired accuracy. In particular, it should be possible to compute (in a finite length of time!) the nearest integer to b.

To see just how powerful this assumption is, consider the Fermat numbers: $F_n = 2^{2^n} + 1$. The first few of these are $5, 17, 257, 65537$, and 4294967297. In 1650 Pierre Fermat conjectured that all of them are prime,

5.1. Constructivist Criticism of Classical Mathematics

but in 1732 Euler discovered that $F_5 = 4294967297 = 641 \cdot 6700417$. Subsequently, no other Fermat number has proven to be prime. Nor has anyone been able to prove that all Fermat numbers beyond the first four are composite. Define the sequence α_n by

$$\alpha_n = \begin{cases} 0 & \text{if } F_n \text{ is not prime,} \\ 1 & \text{if } F_n \text{ is prime.} \end{cases}$$

Then $\{\alpha_n | n > 4\}$ is a bounded sequence each term of which is computable (with some difficulty) in a finite amount of time. The least upper bound of $\{\alpha_n | n > 4\}$ is either 0 (if no Fermat number beyond the first four is prime) or 1 (if there is a prime Fermat number F_n for some $n > 4$). Thus the computation of even the nearest integer to this particular least upper bound is equivalent to solving a famous, longstanding, unsolved problem, namely the possible primality of the Fermat numbers. This we are unable to do at all, much less program a computer to do it in a finite amount of time.

Many other unsolved problems can be coded as the least upper bound of a sequence. If the completeness axiom were constructively verifiable, then these problems would all be solvable—in a finite amount of time—by the same computer program, namely the one that finds least upper bounds. We conclude that the completeness of the reals is almost certainly not computationally verifiable.

To the extent that we agree with this discussion and believe that mathematical statements should be constructively verifiable, we are **constructivists**. This word refers, in particular, to a school of mathematicians going back to Kronecker and Brouwer, and represented today by Bishop, Bridges, and others. On the other hand, if we are willing to tolerate statements without strict computational confirmation, then we are **classical** mathematicians. The issue dividing these groups is fundamental: What can and should mathematics mean?

More non-constructive mathematics

The most straight-forward way to prove the existence of some mathematical object is actually to construct the object (e.g., by a process implicitly programmable on a computer). Proof by contradiction, however, allows us to avoid a direct or affirmative existence proof. Instead we prove that the assumption of the object's non-existence leads to a contradictory conclusion. Assuming that mathematics is consistent (that is, that a contradiction is impossible), we are allowed to conclude from this contradiction that the initial assumption is wrong, and hence the object does not not-exist; rather, by double negation, it exists. Such proofs are sometimes called **pure** existence proofs.

To a constructivist, this method of proof is simply wrong. A pure existence proof provides no way actually to construct or compute the object whose existence is being proved. In general, constructivists avoid proof by contradiction, not just in existence proofs. The consistency of mathematics is unproven, and, as Gödel has shown, is extremely unlikely ever to be proven. Belief in it is a matter of faith, hence so is belief in the validity of proof-by-contradiction.

In this chapter, we choose to adopt a fairly strict constructivist point-of-view. Thus, we insist that all proofs be affirmative, all statements be computationally verifiable, and all objects introduced be finitely constructible. We explore the consequences of these assumptions for the real number system.

Note that a proof by mathematical induction is constructive. This may seem paradoxical because proofs by induction prove an infinite number of things at once. However, to reach any single instance of the conclusion of a proof by induction requires only a finite number of induction steps, hence is a finite proof.

Who cares?

Few mathematicians are practicing constructivists. Of what interest then is constructivism? The typical mathematician (whoever that is) probably does not agree with constructivist principles, not in their strict form at least, but most mathematicians see some validity in the constructivist criticisms of classical mathematics and are sympathetic to the constructivist point of view. Although practicing classical mathematics, the average mathematician is enough of a constructivist to prefer affirmative proofs, computationally verifiable statements, and constructive proofs of existence whenever possible. From this point of view the constructive reals are of quite general interest.

Problems

8) Problem 3 suggests that the reader prove that $\{a(n)\}$ is a convergent sequence by applying the lemma (proved in section 1.4) to the effect that, in a complete field, bounded monotonic sequences converge. Criticize this suggestion from a constructivist point of view.

5.2. The Constructivization of Mathematics

Summary

The fundamental issues of mathematical theory and practice raised by constructivists are the non-computational and hence unverifiable nature of some parts of modern mathematics, including pure existence proofs, arguments based on double negation, and the excessive pursuit of abstraction. The rest of this chapter puts forward the positive side of constructivism by presenting the theory of the real numbers formulated by Errett Bishop, the foremost expositor of constructivism in the second half of the 20th century.

5.2 The Constructivization of Mathematics

A clash of philosophies

What is mathematics about? Platonism, the oldest answer to this, the fundamental question of the philosophy of mathematics, holds that mathematics is about some thing, that mathematics is real—even though you don't see numbers, equations, functions, and other mathematical objects around us in the material real world. Platonism (named after the ancient Greek philosopher who was a major force in founding Western philosophy) maintains that mathematical concepts and other abstractions are real in the sense that one can investigate them (mentally at least, if not physically), draw conclusions that link them together in a web of results and consequences, and therefore know them. A Platonist believes that mathematics is out there somewhere waiting for us to find it. This is encapsulated in the Platonist preference for saying that "mathematics is discovered" (rather than "invented"). Platonism is probably the most influential of all mathematical philosophies. It has immense appeal to those who work with and use mathematics. A naive Platonism is the working philosophy of many mathematicians.

One consequence of the belief that mathematics is out there waiting to be discovered is the belief that there is a 'truth of the matter' regarding mathematical conjectures. Platonists (and many other mathematicians who might not label themselves as such) believe, for example, that the statement that there are no prime Fermat numbers F_n for $n \geq 5$, is definitely either true or false, that is, that either all Fermat numbers F_n for $n \geq 5$ are composite, or some one of them is prime. The same dichotomous belief applies to many other simple, declarative mathematical statements; a faith that they *must* be either true or false—even if at present we don't know which.

On the basis of this belief, proof by contradiction and, in particular, pure existence proofs make perfect sense. If a statement must be true or false and you can prove that it is not false, then it has to be true. If there is always a 'truth of the matter,' then it makes sense to base a logic on this circumstance, i.e., the idea that statements are either true or false. Thus one is led to the classical, Aristotelian logic used by most mathematicians.

Constructivists, in contrast, base their logic, not on truth, but on proof. To appreciate how constructive mathematics works, we first learn its logic. We begin by explaining the meaning of classical logical terms.

Classical logic

Here are the rules of classical, Aristotelian, truth-based logic.

(1) **Conjunction.** Let P and Q be statements. A compound statement of the form "P and Q" (in symbols, $P \wedge Q$) means that statement P is true and statement Q is true.

(2) **Disjunction.** Let P and Q be statements. A compound statement of the form "P or Q" (in symbols, $P \vee Q$) means that either statement P is true, or statement Q is true, or both.

(3) **Implication.** Let P and Q be statements. A compound statement of the form "P implies Q" (in symbols, $P \Rightarrow Q$) means that if statement P is true, then so is statement Q.

(4) **Negation.** Let P be a statement. The statement "not P" (in symbols: $\neg P$) means that statement P is false.

(5) **Universal Statements.** Let $Q(x)$ be a statement about an object x. A statement of the form "For all x, Q is true of x" (in symbols, $\forall x Q(x)$), means that given any x, where x comes from some previously understood universal set, $Q(x)$ is true.

(6) **Existence Statements.** Let $Q(x)$ be a statement about an object x. A statement of the form "There is an x, such that Q is true of x" (in symbols, $\exists x Q(x)$), means that there is some x in our universal set such that $Q(x)$ is true.

5.2. The Constructivization of Mathematics

Problems

1) Let A and B stand for statements for which there is a 'truth of the matter.' For each of these deductive arguments explain why it is *valid*, that is, why, if the premises of the argument are true, then the conclusion must be true.

Modus ponens: $A \Rightarrow B$
$\underline{A}$
$\therefore B$

Modus tollens: $A \Rightarrow B$
$\underline{\neg B}$
$\therefore \neg A$

Disjunction: $A \vee B$
$\underline{\neg A}$
$\therefore B$

Contrapositive proof: $\underline{A \Rightarrow B}$
$\therefore \neg B \Rightarrow \neg A$

De Morgan's laws: $\underline{\neg(A \wedge B)}$
$\therefore \neg A \vee \neg B$

$\underline{\neg A \vee \neg B}$
$\therefore \neg(A \wedge B)$

$\underline{\neg(A \vee B)}$
$\therefore \neg A \wedge \neg B$

$\underline{\neg A \wedge \neg B}$
$\therefore \neg(A \vee B)$

(*Examples*: These problems can be done using truth tables. One can also simply talk them through.

Transitivity: $A \Rightarrow B$
$B \Rightarrow C$
$\overline{\therefore A \Rightarrow C}$

For transitivity, the only way that $A \Rightarrow C$ can be false, is if A is true but C is false. There are two cases, depending on whether B is true or false. If B is true, then the second premise, $B \Rightarrow C$, is false; if B is false, then the first premise, $A \Rightarrow B$, is false. Thus the only way the conclusion can be false is if one of the premises is false. This proves that the argument is valid.

Double negation: $\underline{\neg\neg A}$
$\therefore A$

For double negation, supposing that $\neg\neg A$ is true, then it is false that A is false. Since there are only two alternatives in classical logic, A must be true.)

2) Let $A(x)$ stand for a statement about an object x. Explain why each of these classical arguments involving negation is valid.

(a) $\dfrac{\neg(\forall x)A(x)}{\therefore (\exists x)\neg A(x)}$

(b) $\dfrac{(\exists x)\neg A(x)}{\therefore \neg(\forall x)A(x)}$

(c) $\dfrac{\neg(\exists x)A(x)}{\therefore (\forall x)\neg A(x)}$

(d) $\dfrac{(\forall x)\neg A(x)}{\therefore \neg(\exists x)A(x)}$

(*Example*: In (d), we are given that no matter what x is chosen, $A(x)$ is false. Therefore there cannot be an x for which $A(x)$ is true. This is the desired conclusion and the argument is valid.)

Constructive logic

Constructive logic is proof-based. The symbols used ($\vee, \wedge, \Rightarrow, \neg, \forall$, and $\exists$) are the same as for classical logic but their meaning has changed.

(1) **Conjunction.** Let P and Q be statements. A compound statement of the form "P and Q" means that we have a proof of statement P and a proof of statement Q.

(2) **Disjunction.** Let P and Q be statements. A compound statement of the form "P or Q" means that we have a proof of statement P, or a proof of statement Q, and *we know which*.

(3) **Implication.** Let P and Q be statements. A compound statement of the form "P implies Q" means that we have a way to convert a proof of statement P into a proof of statement Q.

(4) **Negation.** Let P be a statement. The statement "not P" means that we have a proof that P implies a contradiction. A convenient contradiction, often used in this context, is $(0 = 1)$. Thus "not P" can be taken to mean $P \Rightarrow (0 = 1)$, which in turn means that if we have a proof of P, then we can also prove that $0 = 1$.

5.2. The Constructivization of Mathematics

(5) **Universal Statements.** Let $Q(x)$ be a statement about an object x. A statement of the form "For all x, $Q(x)$", means that given any x from some previously understood universal set, we have a purely mechanical procedure that will produce a proof of $Q(x)$.

(6) **Existence Statements.** Let $Q(x)$ be a statement about an object x. A statement of the form "There is an x, such that $Q(x)$", means that we have a proof of $Q(x)$ for a specific x that can be explicitly constructed.

Rules (1), (3), and (5) are similar to the classical rules, although different in meaning; rules (2), (4), and (6) are drastically different. The constructive rule for existence, for example, is much more strict than the classical. Also, in classical mathematics one can prove the truth of a statement of the form "$P \vee Q$" without proving either P or Q.

Problems

3) How is it possible in classical mathematics to prove a statement of the form $(P \vee Q)$ without proving either P or Q?

4) Reconsider each of the arguments in problem 1 from the point of view of constructive logic. Which are valid now?

   (*Examples*: Transitivity is as straight-forward for constructive logic as for classical logic. The premise $A \Rightarrow B$ means that a proof of A can be turned into a proof of B (by some mechanism). Likewise, from the second premise, we learn that a proof of B can be turned into a proof of C. So, if we have a proof of A, we use the first mechanism to turn it into a proof of B, and then use the second mechanism to turn that proof into a proof of C. Thus we have the machinery to turn a proof of A into a proof of C.

   Double negation is different. The premise, $\neg\neg A$, means $((A \Rightarrow (0 = 1)) \Rightarrow (0 = 1))$. In words, if we are given that a proof of A can be turned into a proof of a contradiction, then that proof, in turn, can be turned into a proof of a contradiction. However there is no way to prove A itself. Double negation is not a valid constructive argument.

5) Reconsider each of the arguments in problem 2 from the point of view of constructive logic. Which are valid now?

   (*Example*: Consider (d). The premise, $(\forall x)\neg A(x)$, can be understood as $(\forall x)(A(x) \Rightarrow (0 = 1))$, meaning that, no matter what x is, a proof of $A(x)$ can be turned into a proof of a contradiction. Turning to the conclusion, suppose we have a proof of $(\exists x)A(x)$, i.e., we can construct an x for which we can prove $A(x)$. Then, as we have just explained, the premise implies that that proof can be turned into a proof of a contradiction. In other words, $((\exists x)A(x)) \Rightarrow (0 = 1)$ or $\neg(\exists x)A(x)$. This is a valid constructive argument.

Constructive sets

In classical mathematics, a set S is defined by a property P expressed in words or symbols. A set S is the collection of all objects x that have the property P, i.e., $S = \{x \mid P(x) \text{ is true}\}$. In constructive mathematics, a set is not defined by a property, but by a construction. Rather than being the set of objects that possess a property, a set is the totality of objects built according to the rules of the construction. To look at this another way, a set S is given by a computer program and S is the collection of all possible outputs of this program.

Every set, in addition, comes with a notion of equality, an equivalence relation. This notion of equality must also pass the test of computational verifiability; there must be a finite, mechanical procedure for deciding whether two objects in the set are equal.

For example, the rationals $\mathbb{Q}$ is the collection of all symbols p/q where p and q are integers and q is not zero. We can think of these symbols as being built by a machine whose inputs are p and q. The machine must be programmed to reject the input $q = 0$, and for equality we have $p/q = u/v$ if and only if $pv = qu$. Clearly a computer can be programmed to convert the first equality into the second for verification in the set of integers.

Constructive functions

Constructivists distinguish a function on a set S from a more primitive form of correspondence called an operation. Here is the definition.

Definition. An **operation** from a set A to a set B is a rule f that assigns an element $f(x)$ of B to each element x of A. The rule must be implementable by a finite mechanical procedure. The set A is called the **domain** of the operation f.

As with all sets, the sets A and B come with a notion of equality. If f satisfies the condition that

$$f(x_1) = f(x_2), \qquad \text{— equality in } B$$

whenever

$$x_1 = x_2, \qquad \text{— equality in } A$$

then f is called a **function**.

Problems
---

6) Give an example of an operation on $\mathbb{Q}$ that is not a function.

5.3 The Definition of the Constructive Reals

Regular sequences and equality

In Cantor's construction, a real number is named by a Cauchy sequence of rationals. The same approach is used by Bishop for the constructive reals, except that a constructive real number is named by a type of Cauchy sequence with a constructive twist.

Definition. A **constructive real number** is a sequence $\mathbf{x} = \{x_n\}$ of rational numbers such that

$$|x_n - x_m| < 1/m + 1/n,$$

for all positive integers m and n. Such a sequence $\{x_n\}$ is also called a **regular sequence**. The set of all constructive reals is notated $\mathbb{K}$.

The definition of regular sequence $\{x_n\}$ states explicitly how fast the terms of the sequence approach each other. An elementary example is a constant sequence, which clearly satisfies the condition. A less trivial example is the sequence $\{a_n\}$ defined in section 5.1 (see problems 1 and 2 below).

Every constructive set is required to be accompanied by a constructively defined notion of equality. Therefore, we immediately supply the following:

Definition. Two constructive reals $\mathbf{x} = \{x_n\}$ and $\mathbf{y} = \{y_n\}$ are **equal** if

$$|x_n - y_n| \leq \frac{2}{n},$$

for all positive integers n.

This definition ensures that if the sequences $\mathbf{x}$ and $\mathbf{y}$ converge, they will converge to the same limit. The definition also forces the terms of $\mathbf{x}$ and $\mathbf{y}$ to approach each other at a definite rate. A limit-like form of this definition, more convenient for some purposes, is given by the following lemma.

Lemma. *Two constructive real numbers $\mathbf{x} = \{x_n\}$ and $\mathbf{y} = \{y_n\}$ are equal if and only if for every positive integer m there is a positive integer $N(m)$ such that for $n > N(m)$,*

$$|x_n - y_n| \leq \frac{1}{m}.$$

Proof. The integer $N(m)$ is so written as to emphasize that it depends on m. Like all constructive functions, $N(m)$ must be computable in a finite number of steps. Suppose that $\mathbf{x}$ and $\mathbf{y}$ are equal according to the definition of equality. In other words:

$$|x_n - y_n| \leq 2/n.$$

To prove the lemma, we must explain how to compute, for each positive integer m, a positive integer $N(m)$ so that for $n > N(m)$, $|x_n - y_n| \leq 1/m$. After some preliminary thought (!), I choose to set $N(m)$ equal to $2m$. This clearly renders $N(m)$ computable in a finite number of steps. Then if $n > N(m) = 2m$, we have

$$\begin{aligned}|x_n - y_n| &\leq 2/n && \text{— by hypothesis} \\ &< 2/2m = 1/m. && \text{— because } n > 2m\end{aligned}$$

This is the desired conclusion.

Conversely, suppose that $\mathbf{x}$ and $\mathbf{y}$ satisfy the condition of the lemma. We must establish that $\mathbf{x} = \mathbf{y}$. Let n be any positive integer. Likewise, let m be another positive integer, and let r be any integer such that $r > m$ and $r > N(m)$. Then:

5.3. The Definition of the Constructive Reals

$$|x_n - y_n| = |x_n - x_r| + |x_r - y_r| + |y_r - y_n| \quad \text{— triangle inequality}$$

$$= \frac{1}{n} + \frac{1}{r} + \frac{1}{m} + \frac{1}{r} + \frac{1}{n} \quad \text{— hypothesis and by definition of constructive real numbers}$$

$$= \frac{2}{n} + \frac{2}{r} + \frac{1}{m} \quad \text{— simplification}$$

$$< \frac{2}{n} + \frac{3}{m}. \quad \text{— since } r > m.$$

Since the last inequality is true for all positive integers m, it follows that

$$|x_n - y_n| \leq 2/n.$$

(See problem 3.) This is what we wanted to prove. □

Constructive proofs

Reading a constructive proof is pure pleasure. Everything is straightforward. If a certain quantity is needed, why the proof says just how to calculate it from quantities already at hand. Properties of all computed quantities are directly verified by more computation and everything works out.

Finding a constructive proof is another thing entirely. It takes a lot of thought (usually working backwards) to line up these quantities so that everything works out so nicely.

Now we can prove that equality has the properties expected of it.

Theorem 5.3.1. *Equality of the constructive real numbers is an equivalence relation.*

Problems

1) Prove that the sequence $\{a_n\}$ defined inductively by $a_1 = 2$ and

$$a_n = \frac{1}{2}\left(a_{n-1} + \frac{2}{a_{n-1}}\right),$$

satisfies $a_n a_{n+1} > 2$ for all $n \geq 1$. Prove also that

$$a_n - a_{n+1} \leq (a_{n-1} - a_n)/2.$$

2) Prove that the sequence $\{a_n\}$ of problem 1 is a regular sequence, hence defines a constructive real.

   (*Hint*: Prove and use the fact that $1/2^{n-1} \leq 1/n$.)

3) Let x and c be *rational* numbers. Prove constructively using only properties of rationals that if

$$x < c + 1/m,$$

for all positive integers m, then $x \leq c$.

   (*Hint*: You may not use proof by contradiction. Instead let $x = p/q$ and $c = a/b$ and use familiar properties of rational numbers and integers.)

4) Prove Theorem 5.3.1.

5) Prove constructively that the rationals $\mathbb{Q}$ can be embedded in $\mathbb{K}$ by the function $x \to \{x\}$ (a constant sequence).

Arithmetic for the constructive reals

In order to give constructive definitions of addition and multiplication we need these quantities:

Definition. Let $\mathbf{x} = \{x_n\}$ be a constructive real number. The rational number x_n is called the **nth rational approximation** to $\mathbf{x}$. The integer $K(\mathbf{x}) = [|x_1| + 2]$ is called the **canonical bound** for $\mathbf{x}$.

Note that $K(\mathbf{x})$ satisfies $|x_n| < K(\mathbf{x})$ for all positive n (see problem 6), so that $K(\mathbf{x})$ is an upper bound for the sequence $\{x_n\}$. Also, observe that the mapping $\mathbf{x} \to x_n$ is an operation from $\mathbb{K}$ to $\mathbb{Q}$, but not a function. Now we can define addition and multiplication.

Definition. Let $\mathbf{x}$ and $\mathbf{y}$ be constructive real numbers. Let $k = \max(K(\mathbf{x}), K(\mathbf{y}))$. The **sum** of $\mathbf{x}$ and $\mathbf{y}$ is defined to be the constructive real $\mathbf{x} + \mathbf{y}$ given by

$$\mathbf{x} + \mathbf{y} = \{(x+y)_n\} = \{x_{2n} + y_{2n}\}.$$

5.3. The Definition of the Constructive Reals

The **product xy** is given by

$$\mathbf{xy} = \{(xy)_n\} = \{x_{2kn} y_{2kn}\},$$

and the **negation** $-\mathbf{x}$ of $\mathbf{x}$ is defined by $-\mathbf{x} = \{-x_n\}$.

These definitions are set up so that the sum and product of regular sequences are regular.

Theorem 5.3.2. *The sequences* $\mathbf{x} + \mathbf{y}, \mathbf{xy}$ *and* $-\mathbf{x}$ *defined above are constructive real numbers. Furthermore, the operations* $(\mathbf{x}, \mathbf{y}) \to \mathbf{x}+\mathbf{y}, (\mathbf{x}, \mathbf{y}) \to \mathbf{xy},$ *and* $\mathbf{x} \to -\mathbf{x}$ *are well-defined, i.e. they are functions.*

Now we can prove

Theorem 5.3.3. *Addition and multiplication of constructive reals satisfy the commutative, associative, and distributive laws. The constant sequences $\{0\}$ and $\{1\}$ act as additive and multiplicative identities, respectively, and $-\mathbf{x}$ is an additive inverse.*

Thus $\mathbb{K}$ satisfies all the field axioms except the existence of multiplicative inverses, a complicated issue that must be postponed until order is introduced.

Problems

6) Prove constructively that $K(\mathbf{x}) > |x_n|$, where $K(x)$ is the canonical bound of $\mathbf{x}$.

7) Prove that the sum and product of constructive reals are constructive reals.

8) Complete the proof of Theorem 5.3.2 by proving that addition and multiplication are functions of their arguments.

9) Prove Theorem 5.3.3.

10) Prove that the constructive real $\mathbf{x} = \{a_n\}$, from problem 1, satisfies $\mathbf{x}^2 = 2$.

11) Define $|\mathbf{x}|$ and $\max\{\mathbf{x}, \mathbf{y}\}$ for constructive reals by

$$|\mathbf{x}| = \{|\mathbf{x}|_n\} = \{|x_n|\},$$
$$\max\{\mathbf{x}, \mathbf{y}\} = \{\max\{x_n, y_n\}\}.$$

Prove that these operations are functions.

5.4 The Geometry of the Constructive Reals

The ordering of the constructive real line is complicated by the requirements of constructive logic. The classical definition of order links $\geq$ and $>$. Thus $>$ is defined as $\geq$ but *not* $=$. As constructive mathematicians, we want to avoid negative definitions for they lead to situations which require arguments by double negation. We could define $\geq$ as $>$ *or* $=$, but this leads to different difficulties due to the strict constructive interpretation of disjunction. Bishop and Bridges [H1] solve this problem by decoupling ordinary inequality ($\geq$) and strict inequality ($>$), that is, giving them separate definitions.

Definition. A constructive real number **x** is positive if there is a positive integer n such that the rational number x_n satisfies

$$x_n > 1/n.$$

The set of positive constructive reals is written $\mathbb{K}^>$. Given two constructive reals, **x** and **y**, we write $\mathbf{x} > \mathbf{y}$, if $\mathbf{x} - \mathbf{y}$ is positive.

A constructive real number **x** is non-negative if

$$x_n \geq -1/n,$$

for all positive integers n. The set of non-negative constructive reals is written $\mathbb{K}^\geq$. Given two constructive reals, **x** and **y**, we write $\mathbf{x} \geq \mathbf{y}$, if $\mathbf{x} - \mathbf{y}$ is non-negative.

An element of $\mathbb{K}^>$ is not merely a real number, say **x**, but **x** together with an integer n such that x_n is greater than $1/n$. Since there may be many such n for a given **x**, a single constructive real **x** can give rise to many positive constructive reals. However, if we adopt the same notion of equality for the set $\mathbb{K}^>$ as for the set $\mathbb{K}$, then all these positive reals are equal. Similar remarks are valid for $\mathbb{K}^\geq$.

As with equality, it is useful to formulate limit-like definitions of the order relations.

Lemma. *A constructive real number* **x** *is positive, if and only if there is a positive integer N such that $x_m \geq 1/N$ for all positive integers $m > N$.*

A constructive real number **x** *is non-negative if and only if for each positive integer n there is a positive integer $N(n)$ such that $x_m \geq -1/n$ for all $m > N(n)$.*

Problems

1) Compare the constructive definition of the positive reals with the definition used in Cantor's construction (section 2.1). Is there any difference?

2) For the non-negative Cantor reals formulate an appropriate definition and compare it with the constructive definition.

3) Verify that two positive reals **x** and **y** with the same rational approximations, i.e. $x_i = y_i$ for all i, but a different index (such that $x_n > 1/n$ or $y_m > 1/m$, $n \neq m$) are equal as constructive reals.

4) Prove that a constructive real **x** is positive if and only if there is a positive integer N such that $x_m \geq 1/N$ for all integers $m > N$.

   (*Hint*: If **x** is positive, let n be such that $x_n > 1/n$. Now $x_n - 1/n$ is a rational, say p/q. Using the definition of constructive real, show that $x_m \geq 1/2q$ for all $m > 2q$. In other words, we can take $N = 2q$. The converse is easier.)

5) Complete the proof of the lemma by proving that a constructive real **x** is non-negative if and only if for each positive integer n there is a positive integer $N(n)$ such that $x_m \geq -1/n$ for all integers $m > N(n)$.

   (*Hint*: If **x** is non-negative, show that $x_m \geq -1/n$ for all $m > n$. In other words, take $N(n) = n$.)

6) Prove that $|\mathbf{x}|$ is non-negative for a constructive real **x**. (See problem 11 in section 5.3)

Standard properties of order

The alternative characterizations of positive and non-negative constructive reals make possible fairly routine proofs of the next two theorems.

Theorem 5.4.1. *The set of positive constructive reals is closed under addition and multiplication. The set of non-negative constructive reals is also closed under addition and multiplication.*

Theorem 5.4.2. *The order relations $>$ and $\geq$ have the properties:*

(a) Transitivity:

$$\text{if } x > y \text{ and } y > z, \text{ then } x > z,$$
$$\text{if } x \leq y \text{ and } y \geq z, \text{ then } x \geq z.$$

(b) Anti-symmetry:

$$\text{if } x \geq y \text{ and } y \geq x, \text{ then } x = y.$$

These seemingly normal results conceal how complicated the ordering of the constructive reals actually is. For example, we can prove "if $x > y$ or $x = y$, then $x \geq y$" (see problem 8); but it is not true, conversely, that if $x \geq y$ then "either $x > y$ or $x = y$". Here is an example. Let β_n be as defined as follows:

$$\beta_n = \begin{cases} 0 & \text{if every even number between 6 and } n \\ & \text{can be expressed as the sum of two odd primes,} \\ 1/p & \text{if } p \text{ is the smallest even number between 6 and } n \text{ that} \\ & \text{cannot be expressed as the sum of two odd primes.} \end{cases}$$

The sequence $\{\beta_n\}$ defines a constructive real number $\mathbf{b}$ that is non-negative (see Problem 7). According to the famous Goldbach conjecture, every even number greater than 4 is expressible as the sum of two odd primes. If this is so, then $\mathbf{b} = 0$. As yet no one has been able to prove or disprove this, so, we are unable to prove $\mathbf{b} = 0$, and unable to prove $\mathbf{b} > 0$. Thus while $\mathbf{b} \geq 0$, it is untrue constructively that $\mathbf{b} = 0$ or $\mathbf{b} > 0$. Therefore the constructive reals do not satisfy the trichotomy law.

Problems

7) Prove that $\{\beta_n\}$ defines a non-negative constructive real.

8) Prove that if $x > y$ or $x = y$ then $x \geq y$.

9) Prove that $|x + y| \leq |x| + |y|$ for constructive reals.

10) Prove that $\max\{x, y\} \geq x$ and $\max\{x, y\} \geq y$.

11) Prove that the rationals $\mathbb{Q}$ are dense in the constructive reals.

12) Prove that the nth approximation x_n to a constructive real x satisfies

$$|x - x_n| \leq 1/n.$$

(*Hint*: Look at the mth approximation to the constructive real $1/n - |x - x_n|$.)

5.4. The Geometry of the Constructive Reals

13) Examine the proof of the trichotomy law for the Cantor reals in Chapter 2. Criticize this proof from a constructive point-of-view.

14) Let **x** and **y** be constructive reals. Prove: if $\mathbf{x} + \mathbf{y} > 0$ then either $\mathbf{x} > 0$ or $\mathbf{y} > 0$.

15) Let **x** and **y** be constructive reals. If $\mathbf{x} > \mathbf{y}$ prove that for any constructive real **z** either $\mathbf{z} \leq \mathbf{x}$ or $\mathbf{y} \leq \mathbf{z}$.

(*Note*: Problems 14 and 15 are dichotomy laws!)

Multiplicative inverses

So far we have not proven the existence of multiplicative inverses. This defect can be remedied as follows.

Lemma. *Let* **x** *be a non-zero constructive real number. Then there is a positive integer N such that $|x_m| \geq 1/N$ for $m \leq N$.*

Theorem 5.4.3. *Let* **x** *be a non-zero constructive real number and let N be the integer produced in the previous lemma. Then the sequence $\{y_n\}$ defined by*

$$y_n = \begin{cases} \frac{1}{x_{N^3}} & \text{if } n < N \\ \frac{1}{x_{nN^2}} & \text{if } n \geq N \end{cases}$$

is a constructive real, **y** *that is a multiplicative inverse for* **x**.

This completes the proof of the theorem:

Theorem 5.4.4. *The constructive reals $\mathbb{K}$ are a field.*

Problems

16) Prove the lemma above.

(*Hint*: Show that $|\mathbf{x}| = \{|x_n|\}$ is a well-defined function of the constructive real **x** and that $|\mathbf{x}| \geq 0$. Therefore, if $x \neq 0$, $|\mathbf{x}| > 0$.)

17) Prove Theorem 5.4.3.

(*Hint*: To prove that **y** is a constructive real number, prove and use the fact that $|y_n| \leq N$ for all n. To prove that $\mathbf{xy} = 1$, examine the nth approximation to $|\mathbf{xy} - 1|$. The details are complicated. See Bishop and Bridges [H1].)

Summary

The constructive reals have the algebra of a field and the geometry of linear order. The linear order is defective in comparison with the ordering of the classical reals in that it does not satisfy the trichotomy law, although it satisfies several dichotomy laws. This is an essential aspect of the constructive reals, however, an expression of the constructivist philosophy of mathematics. The constructivists argue that the trichotomy law is not computationally verifiable, therefore cannot be a property of the reals.

5.5 Completeness of the Constructive Reals

Are the constructive reals complete?

After the constructive critique of completeness presented earlier, it may be surprising to learn that there are positive results on completeness for the constructive reals.

First, consider Cauchy completeness. Here are the constructive definitions:

Definition. A sequence $\{\mathbf{x}_n\}$ of constructive real numbers **converges** to a limit **b** if for each positive integer k there is a positive integer $N(k)$ such that
$$|\mathbf{x}_n - \mathbf{b}| \leq 1/k, \qquad \text{for } n \geq N(k).$$

A sequence $\{\mathbf{x}_n\}$ is **Cauchy** if for each positive integer k there is a positive integer $M(k)$ such that
$$|\mathbf{x}_m - \mathbf{x}_n| \leq 1/k, \qquad \text{for } m, n \geq M(k).$$

The only difference between these definitions and those given in Chapter 1 is the insistence that $N(k)$ and $M(k)$ be computable. Now we can prove the surprising theorem:

Theorem 5.5.1. *A sequence of constructive reals $\{\mathbf{x}_n\}$ converges if and only if it is Cauchy. Therefore, the constructive reals are Cauchy complete.*

Outline of the Proof: Suppose first that the sequence $\{\mathbf{x}_n\}$ of constructive reals has limit **x**. To prove that $\{\mathbf{x}_n\}$ is Cauchy, let $N(k)$ be given by the definition of limit and set $M(k) = N(2k)$. This shows that $\{\mathbf{x}_n\}$ satisfies the constructive definition of Cauchy sequence.

5.6. The Constructive Calculus

Suppose next that the sequence $\{\mathbf{x}_n\}$ is Cauchy. Let $M(k)$ be given by the definition of Cauchy sequence. Let $N(k) = \max\{3k, M(2k)\}$ and let y_k be the $(2k)$th rational approximation to $\mathbf{x}_{N(k)}$. Then $\{y_k\}$ defines a constructive real $\mathbf{y}$ that is the limit of the sequence $\{\mathbf{x}_n\}$.

Note: This proof explicitly constructs the limit of the sequence $\{\mathbf{x}_n\}$. To complete the details, however, is complicated.

Are the constructive reals *really* complete?

From the attack on the completeness of the reals in section 5.1, it seems clear that the constructivists repudiate order completeness. Therefore, we don't expect that $\mathbb{K}$ is complete.

On the other hand we have established that $\mathbb{K}$ is Cauchy complete and can also prove that it is Archimedean (see problem 1 below). What then of the theorem of Chapter 1 (section 1.4) claiming that a field is complete if, and only if, it is Cauchy complete and Archimedean? Unfortunately some results from Chapter 1 are vulnerable to constructivist criticism (see problem 2). In the world of the constructive reals we have the result, paradoxical for classical mathematicians but not constructivists, that the reals are Cauchy complete and Archimedean, but not order complete.

Problems

1) Formulate a constructive version of the Archimedean property and prove that $\mathbb{K}$ is Archimedean.

2) Criticize from a constructivist point of view the proof (given in section 1.4) that if a field is Cauchy complete and Archimedean it is order complete.

5.6 The Constructive Calculus

Constructive continuity

The usual definition of continuity goes something like this: f is continuous on the interval $[\mathbf{a}, \mathbf{b}]$ if for every $\varepsilon > 0$ and $\mathbf{c}$ in $[\mathbf{a}, \mathbf{b}]$ there is a $\delta > 0$ such that if $|\mathbf{x} - \mathbf{c}| < \delta$ then $|f(\mathbf{x}) - f(\mathbf{c})| < \varepsilon$ (see problem 3 in section 2.4). Converting this to a constructive definition simply requires insisting that δ be explicitly computable from ε and $\mathbf{c}$. Bishop (and other constructivists) go further and also require that δ be independent of **c**. This leads to considerable simplification of the theory. This idea is called **uniform continuity** in classical mathematics. There is no practical difference, Bishop argues, as long as we stick to continuous functions defined on a closed interval since even in classical mathematics such continuous functions are uniformly continuous.

Thus we arrive at this definition:

Definition. A real-valued function f defined on the interval [**a**, **b**] is **continuous** if for every $\varepsilon > 0$ and **c** in [**a**, **b**] there is a $\delta = \omega(\varepsilon) > 0$ such that

$$\text{if } |\mathbf{x} - \mathbf{c}| \le \delta, \text{ then } |f(\mathbf{x}) - f(\mathbf{c})| \le \varepsilon$$

The operation $\omega(\varepsilon)$ is called a **modulus of continuity** for f.

Problems

1) Prove that $f(\mathbf{x}) = \mathbf{x}^2$ is a continuous function on the interval [1, 2].

2) Prove that the sum of two continuous functions is continuous.

3) Prove that the composition of two continuous functions is continuous.

4) Prove classically that a function continuous on a closed and bounded interval is uniformly continuous.

   (*Hint*: Give an indirect proof by bisection.)

Properties of continuous functions

Recall the fundamental theorems about continuous functions discussed in Chapter 2: the intermediate value theorem, the boundedness theorem, and the maximum value theorem. Here we are concerned with their constructive versions.

The intermediate value theorem is almost certainly not computationally verifiable, at least in complete generality. To see this, let $\{\gamma(n)\}$ be any sequence of 1s, 0s, and -1s, and let **b** be the real number

$$\mathbf{b} = \sum_{n=1}^{\infty} \gamma(n) 3^{-n}.$$

5.6. The Constructive Calculus

This infinite series converges in both the classical and the constructive reals. Note that **b** is positive if the first non-zero term in the sequence $\{\gamma(n)\}$ is a $+1$, while **b** is negative if the first non-zero $\gamma(n)$ is -1.

Let f be the continuous function defined for real numbers between 0 and 1 by

$$f(0) = -1, \qquad f(1/3) = f(2/3) = \mathbf{b}, \qquad f(1) = 1,$$

with the stipulation that f is linear on the segments from 0 to $1/3$, from $1/3$ to $2/3$, and from $2/3$ to 1.

By the intermediate value theorem, there is a real number **x** between 0 and 1 such that $f(\mathbf{x}) = 0$. If this has constructive validity, then we should be able to approximate **x** closely enough to determine whether $\mathbf{x} < 1/3$ or $\mathbf{x} > 2/3$; in other words, tell whether **b** is negative, positive, or zero. Thus there should be a computer program that can determine, in a finite amount of time and for any input sequence $\{\gamma(n)\}$, whether its first non-zero entry is 1, or -1.

As an example of how powerful this program would be, consider the particular sequence

$$\gamma(n) = \begin{cases} 1 & \text{if a sequence of 1 billion 8s begins at the } n\text{th place in the decimal expansion of } \pi, \\ -1 & \text{if a sequence of 1 billion 3s begins at the } n\text{th place in the decimal expansion of } \pi, \\ 0 & \text{otherwise.} \end{cases}$$

Applied to γ, our hypothetical program can tell whether the decimal representation of π contains either a sequence of 1 billion 8s or a sequence of 1 billion 3s, and which comes first. The existence of such a program is highly doubtful. Thus the intermediate value theorem almost certainly does not have constructive validity.

In contrast to the intermediate value theorem, the boundedness theorem for continuous functions is provable.

Theorem 5.6.1. *Let $f(\mathbf{x})$ be a continuous real-valued function on the interval $[\mathbf{a}, \mathbf{b}]$. Then there is a constant M such that $|f(\mathbf{x})| \leq M$ for all $\mathbf{x}$ in $[\mathbf{a}, \mathbf{b}]$.*

Problems

5) The method of bisection was used in Chapter 1 to prove the intermediate value theorem. This seems to be a constructive method for finding the zero of a function. Or is it? What is the constructive view of this proof technique?

6) Without using the intermediate value theorem, prove that if **x** is a constructive real and **x** > 0, then there exists a constructive real **y** such that **y** > 0 and $\mathbf{y}^2 = \mathbf{x}$.

7) Prove Theorem 5.6.1.

   (*Hint*: Let $\varepsilon > 0$ be any positive number. Constructively choose real numbers $\mathbf{a}_i$ for $i = 0, 1, \ldots, n$, so that

   $$\mathbf{a} = \mathbf{a}_0 \leq \mathbf{a}_1 \leq \mathbf{a}_2 \leq \cdots \leq \mathbf{a}_n = \mathbf{b},$$

   and

   $$|\mathbf{a}_{i+1} - \mathbf{a}_i| \leq \omega(\varepsilon),$$

   where ω is the modulus of continuity for f. Set $M = \max\{|f(\mathbf{a}_0)|, |f(\mathbf{a}_1)|, \ldots, |f(\mathbf{a}_n)|\} + \varepsilon$. This calculates M explicitly. To complete the proof prove constructively that for any **x** in [**a**, **b**] there is an $\mathbf{a}_i$ such that $|\mathbf{a}_i - \mathbf{x}| \leq \omega(\varepsilon)$. Then verify that $|f(\mathbf{x})| \leq M$.)

8) Decide whether or not a version of the maximum value theorem for continuous functions is provable for the constructive reals. If you formulate such a result, find a proof. Otherwise, attack the theorem from a constructivist point of view.

Differential calculus

It is possible to develop the calculus from a constructive point of view, although some of the details are more complicated than the classical calculus.

Definition. Let f and g be real-valued functions on an interval [**a**, **b**]. Then f is called **differentiable** with **derivative** g if for any $\varepsilon > 0$ there is a $\delta = \delta(\varepsilon)$ such that

$$|f(\mathbf{x}) - f(\mathbf{c}) - g(\mathbf{c})(\mathbf{x} - \mathbf{c})| \leq \varepsilon |\mathbf{x} - \mathbf{c}|,$$

whenever **x** and **c** are in [**a**, **b**] and $|\mathbf{x} - \mathbf{c}| \leq \delta(\varepsilon)$. The operation $\delta(\varepsilon)$ is called a **modulus of differentiability**.

5.6. The Constructive Calculus

As with the constructive definition of continuity, this is classically known as **uniform** differentiability. The more familiar form of the above inequality is

$$\left| \frac{f(\mathbf{x}) - f(\mathbf{c})}{x - c} - g(c) \right| \leq \varepsilon,$$

which shows that $g(\mathbf{c})$ is what classical mathematics calls $f'(\mathbf{c})$.

Problems

9) Prove that a differentiable function is continuous.

10) Let g be a derivative. Prove that g is a continuous function.

11) Prove the product rule for constructive differentiable functions.

12) Prove the chain rule for constructive differentiable functions.

Weak counterexamples

The sequences $\{\alpha(n)\}$, $\{\beta(n)\}$, and $\{\gamma(n)\}$ appearing in this chapter (in connection with discussions of the completeness of the reals, the trichotomy law, and the intermediate value theorem for continuous functions, respectively) are examples of "fugitive sequences." In each case, current mathematical knowledge is unable to resolve the question whether any of their terms are non-zero. Such sequences have been used by constructivists to attack classical results. These attacks are known as **weak counterexamples**.

Problems

13) Weak counterexample arguments have a strong flavor of proof by contradiction. In each case the discussion begins by supposing that the principle concerned *is* computationally valid, and proceeds to deduce conclusions almost certainly not computationally valid. Is this a type of proof by contradiction? Does it represent a compromise of constructivist principle? Or do constructivists get to have their cake and criticize it too?

14) Here is a constructive version of the uncountability of the reals: Let $\{\mathbf{a}_n\}$ be any sequence of reals. If $\mathbf{x}_0$ and $\mathbf{y}_0$ are real numbers such that $\mathbf{x}_0 < \mathbf{y}_0$ then there exists a real number $\mathbf{x}$ such that $\mathbf{x}_0 \leq \mathbf{x} \leq \mathbf{y}_0$ and $\mathbf{x} \neq \mathbf{a}_n$ for all n in $\mathbb{Z}^+$. Prove this result.

15) Bishop and Bridges [H1] write: "There is a paradox growing out of this result [the previous problem] which the reader should resolve. Since every regular sequence of rational numbers [that is every constructive real] can presumably be described by a phrase in the English language, and since the phrases in the English language can be sequentially ordered, the regular sequences of rational numbers [i.e. $\mathbb{K}$] can be sequentially ordered." How is this paradox resolved?

5.7 A Final Word about the Constructive Reals

The constructive school of mathematics (also known as **intuitionism**) was founded by Leopold Kronecker (1823–1891), Jules Henri Poincaré (1854–1912) and Luitzen Brouwer (1881–1967) in the period from 1870 to 1930. Although it attracted a lot of attention, it won few adherents since most mathematicians feared it would lead to the rejection of parts of current mathematics of importance to them. The more recent work of Errett Bishop (1928–1983) showed that much of classical mathematics can be conducted in conformity with constructivist ideals and has reawakened interest in constructivism.

In defense of the constructivists let us point out that all measurements and computations with measurements have only finite precision and are, therefore, rational. By insisting on the primacy of computations with integers and rationals, constructivists are perhaps simply being pragmatic. At the very least, constructive mathematics exposes a whole slew of assumptions upon which classical mathematics is built that formerly lay unquestioned.

Of course, the constructivist vision of mathematical reality just possibly is correct!

References for the constructive reals: In this account of constructivism, we have stayed close to Bishop's ideas as expressed in [H1], but the constructivist movement embraces a remarkable variety of points of view. For a broader treatment, including an account of the history of the movement and an extensive bibliography, see [H3].

6

The Hyperreals

Introduction

The hyperreal number system was invented comparatively recently (in the 1960s). What makes it unusual is that it contains infinitely small numbers: hyperreals so small that they are greater than 0 but less than $1/n$ for all n in $\mathbb{N}$.

Such numbers, called **infinitesimals**, have been around for a long time. Leibniz used them in his development of the calculus, and they are still taught in the form of the symbols dx and dy used in differentiation and integration. Right from the beginning of the calculus (in the 1680s), the use of infinitesimals was criticized, and for years the calculus including infinitesimals was used by scientists and mathematicians without sound logical grounds. When the calculus was finally given a firm foundation (in the mid 1800s), it was using epsilons and deltas, instead of infinitesimals. As a consequence, infinitesimals were further banished from theoretical mathematics. It was a surprise when, in the early 1960s, Abraham Robinson discovered how infinitesimals could be introduced into the real number system and used much like ordinary numbers. Equally surprising was that a development of such interest in the practical realm of the calculus arose in the highly abstract world of mathematical logic.

The hyperreals represent a relatively liberal philosophy of mathematics. Their construction depends on highly non-constructive arguments. In particular, we require an axiom of set theory, the well-ordering principle, which assumes into existence something that cannot be constructed.

6.1 Formal Languages

Suppose you were told,

>"Every sentence that is true for the reals
>is true for the hyperreals."

I hope you wouldn't believe it! How could everything about the reals continue to be true in an enlarged number system including infinitesimals? Along with infinitesimally small numbers, the hyperreals contain infinitely large numbers. The hyperreals are not Archimedean nor complete. So it's not the case that everything true about the reals is true for the hyperreals.

Instead consider the statement,

>"Every sentence in the language $\mathcal{L}$
>that is true for the reals
>is true for the hyperreals."

The new feature is $\mathcal{L}$. In this form, with the right choice of language, the statement might be true. In fact, it is true. This is where mathematical logic enters the picture since $\mathcal{L}$ is one of the formal languages of modern symbolic logic.

Symbols

Every written language uses symbols. Natural languages like English have just two types:

grammatical signs: ; , . : ( ) ...

and

constants: a b c d e f g h i j k l m ... .

In contrast, **formal** languages, languages whose grammatical rules can be written down, have a great variety of symbols:

variables: x_1 x_2 x_3 x_4 x_5 x_6 ...

a b c d e f ...

grammatical signs: ( ) ,

connectives: $\vee$ $\wedge$ $\rightarrow$ $\neg$

quantifiers: $\forall$ $\exists$

constant symbols: 1 -21.5 π $\sqrt{2}$ ...

function symbols: $+$ $-$ sin cos ...

relation symbols: $=$ $<$ $>$ $\leq$ $\geq$ ...

6.1. Formal Languages

The purpose of each symbol type is easily stated. Variables and grammatical signs perform their usual mathematical functions. Variables stand for individuals, members of whatever set we may be studying, usually, in this book, a set of numbers. Grammatical signs (parentheses and comma) are for grouping and listing. The constant, function, and relation symbols stand for particular constants, functions, and relations that arise in whatever universe we are studying. The examples of constants, functions, and relations given above are all numerical, but formal languages can be used to study any kind of set (see the problems below).

It is the connectives and quantifiers ($\vee \wedge \rightarrow \neg \forall \exists$) that give formal languages their unique flavor. They stand for thephrases indicated above: the logical phrases most often used in mathematics—in definitions, theorems and proofs. For example, the trichotomy law written in these symbols, becomes:
$$\forall x((x > 0) \vee (x < 0) \vee (x = 0)),$$
that is,
"For every x, either $x > 0$, or $x < 0$, or $x = 0$.",
while the existence of additive inverses becomes:
$$\forall x \quad \exists y \quad (x + y = 0)$$
"For all x, there exists a y, such that $x + y$ equals 0".

All the formal languages we study use the same variables, grammatical signs, connectives, and quantifiers. Where they differ is in the vocabulary of constant, function, and relation symbols, which are chosen appropriate to what we want to study.

Syntax

Here we give the grammatical rules for formal languages, rules specifying the form that sentences in the language must have in order to have meaning. Before defining 'sentences,' we define 'terms.' Terms are the nouns of formal languages, the parts of sentences that refer to specific objects.

Definition. A **term** is either
    (a) a constant,
    (b) a variable,
or
    (c) $f(t_1, t_2, \ldots, t_n)$ where f is an n-variable function, and $t_1, t_2, \ldots, t_n$ are terms.

The definition specifies that a term is either a constant, a variable (both are noun-like), or the transformation of other terms by a function. As examples, '4', 'π', 'x', and 'sin($x + \pi$)' are terms. The first two are constants, the third is a variable, and the last combines constants and variables using two functions (addition and the sine). The definition of 'term' is recursive, a kind of definition where the word defined is used to define itself. Proving theorems about concepts that are defined recursively uses a special technique: proof by recursion.

Recursive proof

A **recursive definition** (also called an **inductive definition**) refers to itself. A well-designed recursive definition is perfectly valid. It consists of two parts: a base case in which some instances of the use of the word are defined directly, and an inductive case in which previously defined instances of the word define further instances. For 'term' the base case(s) are parts (a) and (b); (c) is the inductive case.

A recursive definition makes possible **recursive proof**, which is much like mathematical induction. A recursive proof includes separate proofs of base case and inductive case. For the inductive case, one assumes that the theorem is true for previously defined cases. Here is an example:

Theorem 6.1.1. *A term contains an even number of parentheses.*

Proof. By recursion.

Base case: A term that is either a constant or a variable has zero parentheses. Zero is an even number.

Inductive case: In this case a term has the form: $f(t_1, t_2, \ldots, t_n)$, where $t_1, t_2, \ldots, t_n$ are terms for which we assume that the theorem has already been proven. Then the number of parentheses is the sum of the numbers of parentheses in t_1, and in t_2, and so forth, plus two additional parentheses. These numbers are all even according to the inductive assumption. So this case follows because the sum of even numbers is even. □

A term by itself is meaningless. Likewise, in natural languages a noun alone, 'chair' or 'flamboyance,' for example, is (relatively) meaningless.

6.1. Formal Languages

The smallest meaningful unit of a formal language is a sentence, defined as follows:

Definition. A **sentence** is either
  (a) $R(t_1, t_2, \ldots, t_n)$ where R is an n-variable relation and $t_1, t_2, \ldots, t_n$ are terms,

or, assuming that F and G are already sentences, then so are:
  (b) $(F \vee G)$
  (c) $(F \wedge G)$
  (d) $(F \to G)$
  (e) $\neg F$

or, if x is a variable and $H(x)$ is a sentence in which x appears, then these are also sentences:
  (f) $\forall x H(x)$
  (g) $\exists x H(x)$

The definition of 'sentence' is recursive. Sentences of type (a) are the *base case*; types (b) through (g) are recursive cases.

Sentences of type (a) are the simple declarative sentences of formal languages. For example, if $I(x)$ is the one variable relation "x is an integer", then the sentence $I(4)$ says "4 is an integer"; or if $R(x, y)$ is the two variable relation "x is greater than y", then $R(1, 2)$ says "1 is greater than 2". Sentences of this type are the simplest grammatical units that have a truth value.

Every sentence is either a simple declarative sentence like the examples just given or a sentence formed from other sentences either by connecting them using 'and', 'or', 'if-then', or 'not', or quantifying them using 'for all' or 'there exists'.

These rules are called the **formation rules** of formal languages and completely describe their **grammar** (or **syntax**).

Problems
---

1) Prove that every sentence contains an even number of parentheses.

2) Show that the length of a term, meaning the number of characters it contains, can be any natural number with three exceptions. What are the exceptions?

3) What are the possible lengths of sentences?

A language for maps

Some examples will clarify all this terminology. Our first example, the language $\mathfrak{M}$, is designed to make statements about a specific map, the map in Figure 6.1.1. In addition to the standard variables, grammatical signs, connectives and quantifiers, we let $\mathfrak{M}$ contain:

constants: S, T, U, V, W, — for the five regions of the map,

and

relation: $N(x, y)$. — for "x and y are neighbors".

Figure 6.1.1. A map.

Problems

4) For the following sentences in $\mathfrak{M}$, decide which type of formula it is according to the formation rules (a)–(g). In addition, determine its truth value.

(a) $N(S, T)$
(b) $N(S, W)$
(c) $(N(S, T) \vee N(S, W))$
(d) $(N(S, T) \vee N(S, W))$
(e) $(N(S, T) \vee N(S, U))$
(f) $(N(S, V) \wedge N(S, W))$
(g) $\forall x (N(x, T) \to N(x, U))$
(h) $\exists x (N(x, S) \wedge N(x, W))$

(*Examples*: Consider the formula: $N(T, U)$. This is formed according to rule (a). It asserts that T and U are neighbors, which is true. The formula: $N(S, U) \wedge N(S, V)$, on the other hand, is formed according to rule (c). It asserts that S is neighbor of both U and V, which is false.)

6.1. Formal Languages

5) For what values of x are these statements true:
   (a) $N(x, V)$
   (b) $\neg N(T, x)$
   (c) $\neg N(x, S) \wedge \neg N(x, T)$
   (d) $\neg N(x, S) \vee \neg N(x, T)$
   (e) $N(x, V) \rightarrow N(x, W)$
   (f) $\forall y N(x, y)$
   (g) $\forall y (N(x, y) \wedge N(y, S))$

   (*Example*: Consider the formula: $\neg N(x, T)$. This sentence says that x is not a neighbor of T. W is the only value of x that makes this sentence true.)

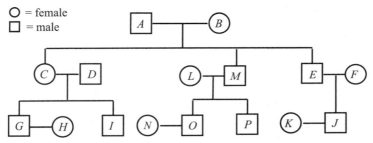

Figure 6.1.2. A genealogical table.

A language for genealogy

The next language, $\mathfrak{G}$, is designed to make statements about a particular genealogical table (see Figure 6.1.2). This example is adopted from [I2]. In addition to the usual variables, grammatical signs, connectives and quantifiers, $\mathfrak{G}$ contains:

   constants: $A, B, C, D, E, F, G, H, I, J, K, L, M, N, O, P$
   — individuals in the tree,

and

   relations: $P(x, y)$ — for "x is parent of y"
   $M(x, y)$ — for "x and y are married"
   $F(x)$ — for "x is female"
   $E(x, y)$ — for "x and y are the same person".

Problems

6) True or False?

    (a) $\forall x \neg M(x, I)$
    (b) $\forall x (P(C, x) \to \neg F(x))$
    (c) $\forall x (P(x, C) \to \neg F(x))$
    (d) $\forall x P(x, O)$
    (e) $\exists x M(x, O)$
    (f) $\exists x P(O, x)$
    (g) $\forall x (\exists y (P(y, x)) \to \neg F(x))$
    (h) $\forall x (\exists y (P(y, x) \vee \neg(x, C)) \to \neg F(x))$

7) Who is x?

    (a) $M(x, G)$
    (b) $(P(A, x) \vee \neg F(x))$
    (c) $\exists y (P(y, x) \vee \exists z P(x, z) \vee \neg F(x))$
    (d) $F(x) \vee \exists y \exists z (P(x, y) \vee P(x, z) \vee (F(y) \to \neg F(z)) \vee (F(z) \to \neg F(y)))$

8) Translate from English to $\mathfrak{G}$.

    (a) "x is a sister of y"
    (b) "x is a grandfather of y"
    (c) "x is a niece of y"
    (d) "x is a first cousin of y"
    (e) "x is a sister-in-law of y"

    (*Example*: "x is the mother of y" can be expressed in $\mathfrak{G}$ by the sentence $P(x, y) \vee F(x)$.)

A language for the integers

The language $\mathfrak{J}$ is designed to make simple statements about the number system $\mathbb{Z}$. In addition to the usual connectives, variables, quantifiers, and grammatical symbols, $\mathfrak{J}$ contains

6.1. Formal Languages

constants:	$-2, -1, 0, 1, 2, 3, \ldots,$	— one for each integer:
functions:	$s(x) = x^2$	— squaring of integers
	$A(x, y) = x + y$	— addition of integers
	$M(x, y) = xy,$	— multiplication of integers

and

relations:	$P(x)$	— for "x is positive"
	$E(x, y).$	— for "x and y are equal"

Note that $\mathfrak{J}$ has an infinite number of symbols.

Problems

9) Determine the truth or falsity of these sentences.

   (a) $P(1)$
   (b) $\neg P(-2)$
   (c) $P(0)$
   (d) $E(A(1, 1), 2)$
   (e) $\forall x (P(s(x)))$
   (f) $\forall x (\exists y E(s(y), x))$
   (g) $\forall x \forall y ((P(x) \vee P(y)) \rightarrow P(A(x, y)))$

10) Express in $\mathfrak{J}$.

   (a) "x is even"
   (b) "x is the additive inverse of y"
   (c) "There is an infinite number of even numbers"
   (d) "x is prime"
   (e) "There is an infinite number of prime numbers"
   (f) "Every even number is the sum of two odd primes"

   (*Example*: The sentence "There are an infinite number of squares" can only be expressed indirectly in $\mathfrak{J}$ because the given functions and relations don't include the relation "there is an infinite number of x".

We must express the idea of infinity using only the ideas of squaring, addition, multiplication, positivity and equality.

The infinitude of the perfect squares can be effectively conveyed if we write a sentence that says that no matter how far out one goes in the integers, there is always another square further on. After some thought, we use the following sentence

$$\forall x \exists y \exists z \exists w (E(s(y), z) \lor E(A(x, w), 0) \lor P(A(z, w)))$$

which says: for every integer x, there is a square ($z = y^2$) that is bigger than x, since $z - x$ is positive (note that $w = -x$).)

Summary

The languages $\mathfrak{M}$, $\mathfrak{G}$ and $\mathfrak{J}$ introduced in this section are **formal languages**. Such languages have rigid grammatical rules that can be written down explicitly, unlike **natural languages**, whose grammars cannot be completely written down and which are constantly evolving. Formal languages are suitable for arguing about mathematical statements; natural languages are suitable for ordinary life.

6.2 A Language for the Hyperreals

We now define the language $\mathfrak{L}$ whose every sentence, if true of the reals, is also true of the hyperreals!

Definition. The language $\mathfrak{L}$ consists of the usual variables, connectives, quantifiers, and grammatical signs, plus the following:

> *constants*: one symbol for every real number,
>
> *functions*: one symbol for every real-valued function of any finite number of real variables,

and

> *relations*: one symbol for every relation on the real numbers of any finite number of variables.

The expressive power of $\mathfrak{L}$ is considerable. For example, all the axioms of the reals (completeness excepted) can be expressed rather easily in $\mathfrak{L}$. The problems explore this.

6.2. A Language for the Hyperreals

Problems

1) Write the field axioms in $\mathfrak{L}$.

   (*Example*: The commutative law of addition expressed in $\mathfrak{L}$ is
   $$\forall x \forall y ((x + y) = (y + x)).$$

   Note: $\mathfrak{L}$ has a symbol for every real number and every real function and relation. We use the same symbols for these numbers, functions, and relations both inside $\mathfrak{L}$ and outside $\mathfrak{L}$. Thus, the plus sign (+) is used for the two-variable function of addition in $\mathfrak{L}$. Likewise, 1 is used for the number one in $\mathfrak{L}$. Strictly speaking $\mathfrak{L}$ uses a *copy* of each symbol from the reals. One would have to use these copies if a situation arose in which one might confuse the symbols of $\mathfrak{L}$ with the reals themselves.)

2) Write the order axioms in $\mathfrak{L}$.

3) For a set of reals S, let R_S be the relation of "membership in S": that is, for a real number r, $R_S(r)$ means "r is a member of the set S". Since $\mathfrak{L}$ contains a symbol for every relation on the real numbers, $\mathfrak{L}$ contains the relation R_S. In $\mathfrak{L}$, write a sentence that says that the set S has an upper bound.

4) For a set of reals S, write a sentence in $\mathfrak{L}$ that says that S has a least upper bound.

5) Can you express the completeness axiom in $\mathfrak{L}$?

   (*Hint*: In $\mathfrak{L}$ one can say "for all numbers x" (i.e. $\forall x$), but one cannot say "for all sets S".)

Structures

Structures add meaning to a formal language by supplying an interpretation for the sentences of that language so that they become true or false. Here's how this is done.

Definition. A **structure** for a language consists of:
  (a) a set S so that each constant symbol in the language corresponds to an element of S
  (b) a set F of functions on S so that each function symbol in the language corresponds to a function in F, and
  (c) a set R of relations on S so that each relation symbol in the language corresponds to a relation in R.

A structure gives meaning to a formal language by specifying things to which its constant, relation, and function symbols can refer. Each of the example languages in section 6.1 was presented along with a typical structure. For $\mathfrak{M}$, the structure is (a) a particular set of five regions for the five constant symbols of $\mathfrak{M}$ (see Figure 6.1.1), (b) *no* functions because the language $\mathfrak{M}$ has no function symbols, and (c) a single relation for the neighbor relation symbol. For $\mathfrak{G}$, the structure is a specific set of sixteen individuals (see Figure 6.1.2) and four familial relations. For $\mathfrak{J}$, the structure is the set of integers plus several functions and relations already defined for integers.

In these examples, the structure we supplied was natural for each language because, in fact, those languages were invented to fit the structures. This is the way it usually is for a formal language. First comes a structure that we intend to study, or more likely have already studied, typically a piece of mathematics or of computer science or of the real world. The formal language is invented to facilitate that study. It is easy to imagine other structures to suit each of our examples: for $\mathfrak{M}$, other maps; for $\mathfrak{G}$, other genealogical tables; for $\mathfrak{J}$, other algebraic systems. There actually is nothing about the language $\mathfrak{M}$ that requires it be used for maps. A structure for $\mathfrak{M}$ could be any set with at least one element (five aren't needed because the constant symbols don't have to refer to different objects) and one relation. Similar remarks apply to the languages $\mathfrak{G}$ and $\mathfrak{J}$.

Turning to $\mathfrak{L}$, its intended structure is the real number system $\mathbb{R}$ including all real functions and relations. Although obvious, this is important enough to state formally.

Theorem 6.2.1. *The real number system $\mathbb{R}$ is a structure for the language $\mathfrak{L}$.*

Although $\mathfrak{L}$ is designed specifically for the reals, like any other formal language it has other structures, the hyperreals for example.

6.2. A Language for the Hyperreals

Definition. A **hyperreal number system** is a structure for the language $\mathcal{L}$ that in addition to all real numbers contains an **infinitesimal** number, a number $*$ such that for all positive integers n, $0 < * < 1/n$.

Problems

6) Find a structure for $\mathfrak{M}$ such that $N(T, U)$, $\neg N(U, V)$, and $\forall x N(x, S)$ are true.

7) Find a structure for $\mathfrak{M}$ such that the sentences in problem 6 are false.

8) Find a structure for $\mathfrak{G}$ such that $\forall x (\neg E(x, M) \rightarrow (P(M, x) \vee F(x)))$ is true.

9) Find a structure for $\mathfrak{G}$ such that $\forall x (\neg E(x, M) \rightarrow \exists y P(x, y))$ and $\forall x \forall y ((\exists z P(x, z) \vee P(y, z)) \rightarrow E(x, y))$ are true

10) Find a structure for the language $\mathfrak{J}$ such that $\forall x \forall y (E(s(x), s(y)) \rightarrow E(x, y))$ is true.

(*Hint*: This is the only sentence that has to be true.)

Summary

We claim that $\mathcal{L}$, the language introduced in this section, is such that all theorems about the reals that can be expressed in $\mathcal{L}$ are also true of the hyperreals. Whether or not this is correct, $\mathcal{L}$ is powerful enough to express the field axioms, and many other statements about number systems. It appears that $\mathcal{L}$ is also chosen for what it cannot express. It seems the completeness axiom of the reals cannot be written in $\mathcal{L}$. This suits the purpose for which $\mathcal{L}$ is intended, for the hyperreals cannot satisfy the completeness axiom.

A structure gives meaning to a formal language. An abstract language, such as $\mathcal{L}$ (which was designed to have the real number system together with all its relations and functions as a structure), can always be applied to other structures. A hyperreal number system is a structure for $\mathcal{L}$ which contains infinitely small numbers.

6.3 Construction of the Hyperreals

Sequences again

The construction of the hyperreals resembles Cantor's construction of the reals in that it uses an equivalence relation applied to a set of sequences. Those sequences then function as the names of hyperreals.

Definition. The set $\mathbb{H}$ of **hyperreals** is the set of all infinite sequences $\{x_n\}$ of real numbers.

The hyperreals uses all possible sequences. A typical hyperreal is a sequence of reals that might look something like this:

$$\mathbf{a} : \{52, -1, 17, 3, -1.009163, 4 - \sqrt{7}, 4.0193(10^{14}), 2\pi, \ldots\}.$$

Thus a hyperreal, as a sequence, might appear quite random. $\mathbb{H}$ also contains highly patterned sequences. In particular, $\mathbb{H}$ contains constant sequences such as

$$\mathbf{b} : \{1, 1, 1, 1, \ldots\},$$

and

$$\mathbf{c} : \{2, 2, 2, 2, \ldots\}.$$

These play the roles of the numbers 1 and 2 in $\mathbb{H}$, and the other reals are embedded in $\mathbb{H}$ this way too.

What about a sequence like

$$\mathbf{d} : \{-2, 3, 2, 6.54321, \pi, 2, 2, 2, 2, 2, 2, 2, 2, 2, \ldots\},$$

which is only slightly more complicated than a constant? For most n, $\mathbf{d}_n = \mathbf{c}_n$. For only four subscripts, 1, 2, 4 and 5, do $\mathbf{d}$ and $\mathbf{c}$ differ. Under these circumstances, we stipulate that

$$\mathbf{c} = \mathbf{d} \text{ (in } \mathbb{H}).$$

This requires introducing an equivalence relation for equality.

What sort of number is the sequence

$$\mathbf{e} : \{1, -1, 1, -1, 1, -1, \ldots\}?$$

If we do what seems natural and square $\mathbf{e}$ we get

$$\mathbf{e}^2 : \{1, 1, 1, 1, 1, 1, \ldots\};$$

so $\mathbf{e}$ is a square root of 1. This suggests that $\mathbf{e}$ is either 1 or -1, but which?

6.3. Construction of the Hyperreals

These problems will be resolved in this way: each place in a hyperreal will not by itself influence the nature of that hyperreal. The sequences **c** and **d** differ at only four subscripts. Four is not enough! We consider two hyperreals different only if they differ on a *big* set of subscripts. Note that a big set is a subset of $\mathbb{N} = \{1, 2, 3, \ldots\}$ since the natural numbers are the numbers used for subscripts.

Which subsets of $\mathbb{N}$ are big? We define 'big' eventually, but for the moment the reader is asked to accept the definition of equality and order that follows, which uses big sets without defining them.

Definition. Two hyperreals $\mathbf{x} = \{x_n\}$ and $\mathbf{y} = \{y_n\}$ are **equal** if $\{n | x_n = y_n\}$ is a big set of natural numbers. A hyperreal $\mathbf{x} = \{x_n\}$ is **positive** if $\{n | x_n > 0\}$ is a big set of natural numbers.

To proceed, let us first agree that finite sets are not big. Only an infinite set can be big, but not even all infinite sets. If all infinite sets were big, we would still be unsure what to do with the sequence **e**. Next let us include among the big sets the infinite sets whose complement is finite, the so-called **cofinite** sets. Further properties of big sets are in the next two lemmas.

Lemma 1. *If equality of hyperreals is an equivalence relation, then the big sets have the property: if A and B are big, then $A \cap B$ is big.*

Proof. Let $\mathbf{x} = \{x_n\}$, $\mathbf{y} = \{y_n\}$, and $\mathbf{z} = \{z_n\}$ be sequences defined as follows:

$$x_n = \begin{cases} 0 & \text{if } n \in A \\ 1 & \text{otherwise}, \end{cases}$$

$$y_n = \begin{cases} 0 & \text{if } n \in A \\ 2 & \text{otherwise}, \end{cases}$$

$$z_n = \begin{cases} y_n & \text{if } n \in B \\ 3 & \text{otherwise}. \end{cases}$$

Notice that $A = \{n | x_n = y_n\}$ and $B = \{n | y_n = z_n\}$. Therefore, if A and B are big, then $\mathbf{x} = \mathbf{y}$ and $\mathbf{y} = \mathbf{z}$. By the transitive law, it follows that $\mathbf{x} = \mathbf{z}$, so that $\{n | x_n = z_n\} = A \cap B$ must be big. $\square$

Lemma 2. *If the hyperreals are linearly ordered, then the big sets have the property: for any subset A of $\mathbb{N}$, either A is big or the complement of A is big.*

Collecting the properties of big sets described so far, together with another, natural property gives the following definition. (The technical term for what we have been calling the big sets is the bizarre word 'ultrafilter'.)

Definition. A collection $\mathfrak{U}$ of subsets of $\mathbb{N}$ is a (free) **ultrafilter** if:

(a) $\mathfrak{U}$ contains the cofinite sets, but no finite set,

(b) if A and B are in $\mathfrak{U}$, then $A \cap B$ is in $\mathfrak{U}$,

(c) if A is in $\mathfrak{U}$ and B contains A, then B is in $\mathfrak{U}$,

(d) for every subset A of $\mathbb{N}$, either A or the complement of A is in $\mathbb{U}$.

We intend to use an ultrafilter as the big sets. Of course it is not clear that any such thing as an ultrafilter exists. A proof appears later (in section 6.7). For the moment, let us make the

Assumption. There exists an ultrafilter.

Then we complete the construction of the hyperreals by the following:

Definition. Choose an ultrafilter $\mathfrak{U}$. The sets in $\mathfrak{U}$ will be called **big**.

Problems

1) Prove Lemma 2.

   (*Hint*: Define $\{y_n\}$ as in the proof of Lemma 1. Use the trichotomy law.)

2) Let $\mathfrak{U}$ be an ultrafilter. If A is a subset of $\mathbb{N}$, prove that it is not possible that both A and the complement of A belong to $\mathfrak{U}$.

3) Prove that equality for hyperreals is an equivalence relation assuming that big sets come from an ultrafilter.

4) Prove that the hyperreals are linearly ordered.

6.3. Construction of the Hyperreals

5) Prove that if $A \cap B$ is a big set, then either A or B is big.

6) If $\mathbf{x} = \{x_n\}$ and $\mathbf{y} = \{y_n\}$ are hyperreals, define their product by

$$\mathbf{xy} = \{x_n y_n\}.$$

Prove the law of integral domains: if $\mathbf{xy} = 0$ then either $\mathbf{x} = 0$ or $\mathbf{y} = 0$.

Summary

The hyperreals $\mathbb{H}$ are defined as the set of all real sequences, subject to a peculiar equivalence relation: two sequences are equal if their entries are equal on a big set of subscripts. A big set, in turn, is a set from an ultrafilter, a collection of subsets of the natural numbers satisfying four properties. We prove later that ultrafilters exist.

The use of an ultrafilter solves most of the difficulties concerning equality and positivity posed by our definition of the hyperreals. For example, we can prove that the hyperreals are linearly ordered (problem 4 above).

One problem remains unresolved: the nature of the hyperreal:

$$\mathbf{e} : \{1, -1, 1, -1, 1, -1, \ldots\}.$$

Since the big sets form an ultrafilter, we know that either the set of even subscripts or the set of odd subscripts is big. Therefore, $\mathbf{e}$ is either -1 or $+1$. So far, so good. However, as we don't know whether a given infinite set is big or not (unless it is cofinite), we have no idea whether it is the evens or the odds that are big. Thus, $\mathbf{e} = 1$ or $\mathbf{e} = -1$, but we still don't know which!

Perhaps this problem only persists because we have assumed the existence of an ultrafilter. Even after the existence of an ultrafilter is proven, however, we still will have no way to settle this question. All known existence proofs for ultrafilters are pure existence proofs. As such, they give no clue to the contents of an ultrafilter. We are doomed to eternal ignorance, never knowing whether $\mathbf{e}$ is plus or minus one.

(More accurately, when proving the existence of an ultrafilter, it is possible to force $\mathbf{e}$ to be either 1 or -1, whichever we feel like at that moment, but there will always remain similar undecidable sequences).

6.4 The Transfer Principle
ℍ is a structure

Our goal in this section is to prove that "Every sentence in the language $\mathcal{L}$ true for the reals, is true for the hyperreals." First, however, we must show that the statements of $\mathcal{L}$ apply to the hyperreals, that is, that the constant, relation, and function symbols of $\mathcal{L}$ refer to the constants, relations, and functions of ℍ. We must show that ℍ is a structure for the language $\mathcal{L}$.

We have already defined equality and positivity, but now every other real function and relation must be defined for the hyperreals. Miraculously there is no need to do this one relation and function at a time. There is a mechanism that carries over to ℍ all the functions and relations of ℝ at once. We take relations first.

Definition. Let R be a one variable relation on the reals; that is, for every real number x, $R(x)$ is a sentence that is either true or false. The extension of R to the hyperreals is denoted *R. For a hyperreal $\mathbf{x} = \{x_n\}$, we define *$R(\mathbf{x})$ as **true** if and only if the set

$$\{n \mid R(x_n) \text{ is true in } \mathbb{R}\}$$

is big. Otherwise, *$R(\mathbf{x})$ is **false**.

Let $R(,)$ be a two variable relation on the reals. The extension of R to the hyperreals is denoted *R. For hyperreals $\mathbf{x}$ and $\mathbf{y}$, we define *$R(\mathbf{x}, \mathbf{y})$ to be **true** if and only if

$$\{n \mid R(x_n, y_n) \text{ is true in } \mathbb{R}\}$$

is a big set. Otherwise, *$R(\mathbf{x}, \mathbf{y})$ is **false**.

Similarly, every n variable relation R on ℝ can be extended to an n-variable relation *R on ℍ.

Among the relations now defined on ℍ, are those that distinguish various types of numbers: the integers, rationals, primes, squares, and so forth. Each is now extended to the hyperreals, so we can distinguish hyperreal rational numbers, hyperreal perfect squares and hyperreal prime numbers. For example, let

$$\mathbf{✪} = \{1, 2, 3, 4, 5, \ldots\}.$$

Then ✪ is a hyperreal *integer* because the predicate $I(x) =$ "x is an integer" is true of ✪ at every place—definitely a big set of subscripts. This makes

6.4. The Transfer Principle

the statement $I(\odot)$ true, by definition, so $\odot$ is a hyperreal integer. To carry this discussion a little bit further, we can't say whether $\odot$ is even or odd, or whether it is prime or not, because we don't know whether the set of even subscripts (where $\odot$ has even entries), or odd subscripts (where $\odot$ has odd entries), or prime subscripts (where the entry in $\odot$ is prime) is big. In other words, we don't know whether any of the predicates

$$E(x) = \text{``}x \text{ is even''},$$
$$O(x) = \text{``}x \text{ is odd''},$$

or

$$PN(x) = \text{``}x \text{ is prime''}$$

are true for $\odot$ at a big set of subscripts. We *can* say that $\odot > 10$ since the set of subscripts of $\odot$ where the predicate $GR(x, 10) = \text{``}x$ is greater than 10" is true *is* a big set. (It is co-finite.) Therefore, by definition, $GR(\odot, 10)$ is true, and $\odot > 10$. In fact, $\odot$ is bigger than every single ordinary integer; $\odot$ is infinite.

The reader may be surprised that we are permitted to define the meaning of 'true' and 'false' for the relations *R on $\mathbb{H}$. There is nothing capricious about this, however. The situation is analogous to that moment when Euler defined the complex exponential (see Chapter 3). Before Euler, there was no value for e^{ix}. Since there was no definition, in principle any formula could have been chosen for e^{ix}. Euler's formula is a good choice because it works. It fits with existing facts about the reals and complexes and so is regarded as correct.

A moment ago none of the relations "x is an integer", "x is even" and so forth were defined for the hyperreals. In principle, we can adopt any definition for these relations. We get to say what they mean, but only a definition that, like Euler's formula, fits with existing facts about the reals will prove useful. It is essential that truth and falsity for the relation *R on the hyperreals agree with the truth and falsity of the original relation R on the reals. We address this question in the course of proving that the hyperreals are a structure for $\mathfrak{L}$.

Next we extend every function f on $\mathbb{R}$ to a function *f on $\mathbb{H}$:

Definition. Let $f(x)$ be a one variable real-valued function defined on $\mathbb{R}$. The extension of f to the hyperreals is denoted *f. For a hyperreal $\mathbf{x} = \{x_1, x_2, x_3, \ldots\}$, we define *$f(\mathbf{x})$ by

$$*f(\mathbf{x}) = \{f(x_1), f(x_2), f(x_3), \ldots\}.$$

Let $f(x, y)$ be a two variable real-valued function defined on the reals. The extension of f to the hyperreals is denoted $*f$. For hyperreals $\mathbf{x}$ and $\mathbf{y}$ we define $*f(\mathbf{x}, \mathbf{y})$ to be the hyperreal

$$*f(\mathbf{x}, \mathbf{y}) = \{f(x_1, y_1), f(x_2, y_2), f(x_3, y_3), \ldots\}.$$

Similarly, any n variable real-valued function f on $\mathbb{R}$ is extended to an n-variable function $*f$ on $\mathbb{H}$.

This definition creates a function on $\mathbb{H}$ for every function defined on $\mathbb{R}$. Among the functions thus defined are those of addition, $A(x, y) = x + y$, and multiplication, $M(x, y) = xy$. We have also defined many other functions, for example, we can now calculate the sine of a hyperreal. Thus, if $\mathbf{O} = \{1, 2, 3, 4, 5, \ldots\}$, then

$$\sin(\mathbf{O}) = \{\sin(1), \sin(2), \sin(3), \sin(4), \sin(5), \ldots\}.$$

It's not clear just what kind of number $\sin(\mathbf{O})$ is, but at least we can say that

$$-1 \leq \sin(\mathbf{O}) \leq 1,$$

since this inequality is true for all the entries in $\sin(\mathbf{O})$.

In connection with the last two definitions it is important to prove:

Theorem 6.4.1. *The relations $*R$ and the functions $*f$ are well-defined.*

Problems

1) Let R be a one variable relation on the reals. Prove that $*R$ is well-defined by verifying that if $*R(\mathbf{x})$ is true and $\mathbf{x} = \mathbf{y}$ then $*R(\mathbf{y})$ is true.

2) Let f be a real-valued function of one variable. Prove that $*f$ is well-defined, i.e., that if $\mathbf{x} = \mathbf{y}$ then $*f(\mathbf{x}) = *f(\mathbf{y})$.

3) Among the functions defined on $\mathbb{R}$ are the two variable functions of addition and multiplication: $A(x, y) = x + y$ and $M(x, y) = xy$. Prove that $*A$ and $*M$ make $\mathbb{H}$ a field.

4) One of the many relations defined on $\mathbb{R}$ is the one place relation of positivity:

$P(x)$ is true if and only if $x > 0$.

Check that the extension $*P$ agrees with the previous definition of positivity on $\mathbb{H}$, and prove that that $*P$ makes $\mathbb{H}$ a linearly ordered field.

6.4. The Transfer Principle

5) Among the relations just introduced for $\mathbb{H}$ are $^*Q(\mathbf{x}) =$ "**x** is a rational number" and $^*PN(\mathbf{x}) =$ "**x** is a prime number". Find examples of rationals and primes in $\mathbb{H}$. Find rationals and primes that are in $\mathbb{H}$ but not in $\mathbb{R}$.

6) Show that $\mathbb{H}$ contains an infinitesimal, that is a number $*$ such that $* > 0$ and $* < r$ for all positive reals r. This proves that $\mathbb{H}$ is a hyperreal number system.
   (*Hint*: Try $* = \{1, 1/2, 1/3, 1/4, \ldots\}$.)

7) Find an infinitely large prime integer in $\mathbb{H}$. Find an infinitely large irrational number.

Embedding again

We embed the reals in the hyperreals just as earlier we embedded the rationals in the reals.

Theorem 6.4.2. *For a real constant x, let x^* be the hyperreal $\{x, x, x, \ldots\}$. Then the transformation $x \to x^*$ is both a field and order isomorphism from $\mathbb{R}$ into $\mathbb{H}$.*

With the reals embedded in the hyperreals, we next verify that the definitions of $^*R$ and $^*f$ on $\mathbb{H}$ agree with their definitions on $\mathbb{R}$. This means proving that these relations and functions act on the embedded reals exactly as they act on the reals themselves. This is accomplished by the two parts of the following theorem.

Theorem 6.4.3. *Let R be a one variable relation on the reals. Let f be a real-valued function of one variable. Then*

(a) *$R(x)$ is true (in the reals), if and only if $^*R(x^*)$ is true (in the hyperreals)*

(b) *$^*f(x^*) = (f(x))^*$.*

This embedding of the reals in the hyperreals goes far beyond the earlier embedding of the rationals in the reals. When the rationals were embedded in the reals, only their arithmetic and order properties went with them. In contrast, the reals are embedded in the hyperreals along with every single relation and function.

To see how remarkable this is consider that rational numbers have denominators. We can define a function $d(q)$ that assigns to each rational its denominator when written in lowest terms. To be unambiguous, we insist that d have positive values. Then, for example, $d(25/75) = 3$ and $d(-12/16) = 4$. But the function d does not extend to the reals in any natural way. Pi has no denominator! Thus there is a function on the rationals that represents a fundamental feature of the rationals but does not extend in any natural way to the reals.

On the other hand, real numbers have decimal expansions. We can define a function $d_1(x)$ that assigns to each real number the tenths place in its decimal expansion. To be unambiguous, we insist that decimal expansions cannot end with an infinite string of 9s. Then, for example,

$$d_1(1/4) = 2, \quad \text{— since } 1/4 = .25$$
$$d_1(\pi) = 1, \quad \text{— since } \pi = 3.1415\ldots$$

and

$$d_1(.1999\ldots) = 2. \quad \text{— since } .1999\cdots = 0.2$$

Due to the extraordinary embedding of the reals in the hyperreals, the function d_1 extends to a function on the hyperreals. A hyperreal has a decimal expansion (more about this later) and, in particular, the function d_1 when extended to the hyperreals, gives the tenths place of that decimal expansion. Furthermore this is not something special about decimal expansions. Every single thing about the reals that can be expressed using functions and relations extends to the hyperreals.

Almost trivially we have:

Theorem 6.4.4. $\mathbb{H}$ *is a hyperreal number system, that is, it is structure for the language $\mathcal{L}$ that contains infinitesimals.*

Since the reals, and all real relations and functions, are embedded in the structure $\mathbb{H}$, from now on we drop the distinction between x and x^*, f and $^*f$, and R and $^*R$, and simply assume that the reals, together with all their functions and relations are inside $\mathbb{H}$.

Problems

8) Prove Theorem 6.4.2.

6.4. The Transfer Principle

9) Let R be a one variable relation on the reals. Prove that $R(x)$ is true (in the reals), if and only if $*R(x*)$ is true (in the hyperreals).

10) Let f be a real-valued function of one variable. Prove that $*f(x*) = (f(x))*$

   (*Note*: Problems 9 and 10 prove Theorem 6.4.3.)

11) Prove Theorem 6.4.4.

12) What is the tenths place in the decimal expansion of $*$? Find the rest of the decimal expansion of $*$.

Łos' theorem

We now prove that

"Every sentence in the language $\mathcal{L}$
that is true for the reals
is true for the hyperreals"

We deduce this from an even more general statement about the hyperreals called Łos' theorem. To state this, requires a second formal language, $\mathcal{L}^*$, one designed to make statements about $\mathbb{H}$, as $\mathcal{L}$ was designed to make statements about $\mathbb{R}$.

The language $\mathcal{L}$ contains constant symbols for all real numbers (e.g., 1, $\sqrt{2}$, and π), but no sentence in $\mathcal{L}$ ever refers to a specific hyperreal that is not also a real. The difference between $\mathcal{L}$ and $\mathcal{L}^*$ is the addition of constant symbols for all these hyperreals.

Definition. The language $\mathcal{L}^*$ consists of all the symbols of the language $\mathcal{L}$ plus a new constant symbol, $\mathbf{k} = \{k_n\}$, for every hyperreal $\mathbf{k}$ that is not an ordinary real.

Łos' theorem describes when a statement in $\mathcal{L}^*$ is true.

Theorem 6.4.5. (*Łos' theorem*). *Let S be a sentence in $\mathcal{L}^*$. Let S_n be the sentence in $\mathcal{L}$ obtained from S by replacing every hyperreal constant symbol $\mathbf{k} = \{k_n\}$ with the real constant symbol k_n. Then, S is true in the structure $\mathbb{H}$ for $\mathcal{L}^*$ if and only if*

$$\{n \mid S_n \text{ is true in the structure } \mathbb{R} \text{ for } \mathcal{L}\}$$

is a big set.

Proof. Let S be a sentence in $\mathcal{L}^*$. The formation rules in section 6.1 divide the sentences of $\mathcal{L}^*$ into seven types, (a) through (g). Each type gets a separate proof. In addition, the proof is recursive, since the formation rules are recursive.

Base case: This is type (a). A sentence S of type (a) is a relation symbol applied to some terms. In this case the conclusion of Łos' theorem simply repeats the definition of truth in $\mathbb{H}$ given at the beginning of this section.

Recursive case: While proving the remaining six cases, we assume the theorem is already proven for the sub-sentences used to construct S. A typical case is a sentence of type (c) for which we give a detailed proof as an example.

Let S be of the form $S = (F \wedge G)$. By the induction hypotheses, we assume the theorem has already been proven for the sub-formulas F and G. Note that $S_n = (F_n \wedge G_n)$. To proceed let

$$A = \{n | S_n \text{ is true}\},$$
$$B = \{n | F_n \text{ is true}\},$$

and

$$C = \{n | G_n \text{ is true}\}.$$

We know, by induction, that

$$F \text{ is true for } \mathbb{H} \text{ if and only if } B \text{ is big},$$

and

$$G \text{ is true for } \mathbb{H} \text{ if and only if } C \text{ is big}.$$

Note that $A = B \cap C$. Now S is true for $\mathbb{H}$ if and only if

F and G are true for $\mathbb{H}$	— logic
if and only if B and C are big	— the induction assumption
if, and only if, A is big.	— property (b) of big sets

This proves Łos theorem for sentences of type (c). □

Corollary. *(The Transfer Principle). Let S be a sentence in $\mathcal{L}$. Then S is true in the structure $\mathbb{R}$ for $\mathcal{L}$ if and only if S is true in the structure $\mathbb{H}$. In other words, every sentence in the language $\mathcal{L}$ that is true for the reals is true for the hyperreals (and vice versa!).*

6.4. The Transfer Principle

Proof. Let S be a sentence in the language $\mathfrak{L}$. Then S is also in the language $\mathfrak{L}^*$. By Łos' Theorem, S is true in $\mathbb{H}$ if and only if $\{n \mid S_n \text{ is true in } \mathbb{R}\}$ is big. However since S does not contain any constants from $\mathbb{H}$ (S is in $\mathfrak{L}$), S_n is the same as S for all n. Thus $\{n \mid S_n \text{ is true in } \mathbb{R}\}$ is either all $\mathbb{N}$ (if S is true in $\mathbb{R}$) or empty (if S is false for $\mathbb{R}$). Therefore, $\{n \mid S_n \text{ is true in } \mathbb{R}\}$ is big if and only if S is true for $\mathbb{R}$. □

This is called the transfer principle because it transfers truth back and forth between $\mathbb{R}$ and $\mathbb{H}$.

Problems

---

14) Prove Łos' Theorem for formulas of types (b), (d), and (e).

15) Prove Łos' Theorem for the quantified formulas (f) and (g).

(*Hint*: If $H(x)$ is a formula of $\mathfrak{L}^*$ that contains the variable x, then $H(x)_n = H_n(x)$, that is, the process of replacing hyperreal constants with real constants doesn't affect the variable x.

Using the transfer principle

Many results that we have proved directly about the hyperreals are easily proven using the transfer principle. The problems explore this.

> ### Proof by transfer principle
>
> The transfer principle makes $\mathfrak{L}$ a powerful tool for proving statements about the reals and the hyperreals. Any statement that can be formulated in $\mathfrak{L}$ is automatically true in both $\mathbb{R}$ and $\mathbb{H}$, as long as it can be shown to be true in one of them.
>
> The power of this method of proof arises from the fact that a statement may mean different things in $\mathbb{R}$ and in $\mathbb{H}$ because the quantifiers have different meanings. In $\mathbb{H}$, quantifiers can and do refer to the infinitesimal and infinite quantities not present in $\mathbb{R}$. Some statements are easier to prove in $\mathbb{H}$, and others easier to prove in $\mathbb{R}$.

Problems

16) Show that the statement "$\mathbb{R}$ is an ordered field" can be expressed in $\mathcal{L}$. Explain how this provides a second proof, beyond problems 3 and 4, that $\mathbb{H}$ is an ordered field.

17) Use the transfer principle to prove that every positive hyperreal has a positive square root.
    (*Hint*: This is true for the reals $\mathbb{R}$. Express it in $\mathcal{L}$.)

18) Use the transfer principle to prove the cancellation laws of addition and multiplication for $\mathbb{H}$.

19) Use the transfer principle to prove the integral domain property: if $\mathbf{xy} = 0$ in $\mathbb{H}$, then either $\mathbf{x} = 0$ or $\mathbf{y} = 0$.

20) Prove that if $\mathbf{p}$ is a hyperreal prime number then there is a hyperreal prime $\mathbf{q}$ such that $\mathbf{q} > \mathbf{p}$.

21) Prove that the trigonometric functions $\sin(x)$ and $\cos(x)$ satisfy the same identities for hyperreals as well as reals. For example, $\sin(\mathbf{x})^2 + \cos(\mathbf{x})^2 = 1$ for any hyperreal $\mathbf{x}$.

Summary

We have established that $\mathbb{H}$ is a structure for the language $\mathcal{L}$, that $\mathbb{H}$ contains infinitesimals, and that any theorem about the reals expressible in $\mathcal{L}$ is true for hyperreals. As a consequence many real concepts, such as square roots, integers, rationals, exponentials, and trigonometric functions, carry over to the hyperreals with their rules, identities, equations, and other properties unchanged. These relations and functions apply to the infinitely small and infinitely large elements of $\mathbb{H}$ as well as the ordinary reals. In the next section we consider these new elements of $\mathbb{H}$ in more detail, piecing together just how they fit together along the hyperreal line.

6.5 The Nature of the Hyperreal Line

Playing with infinitesimals

To compare infinitely small and infinitely large numbers we need the vocabulary provided by the next definition.

6.5. The Nature of the Hyperreal Line

Definition. A hyperreal **x** is **infinite** if $|\mathbf{x}| > r$ for all positive reals r. A hyperreal is **infinitesimal** if $|\mathbf{x}| < r$ for all positive reals r. If **x** is not infinite, then **x** is **finite**.

If hyperreals **x** and **y** are such that $|\mathbf{x} - \mathbf{y}|$ is an infinitesimal, then we say **x** is **infinitely close to y**, and write $\mathbf{x} \approx \mathbf{y}$.

Given hyperreals, **x**, **y**, **z**, if the fraction

$$\frac{|\mathbf{x} - \mathbf{z}|}{|\mathbf{y} - \mathbf{z}|}$$

is an infinitesimal, then we say that **x** is **infinitely closer to z than y** (or that **y** is **infinitely further away from z than x**).

Given positive hyperreals, **x** and **y**, we say **x** is **infinitely smaller than y** (or **y** is **infinitely bigger than x**) if **x** is infinitely closer to zero than **y**, i.e., the fraction $|\mathbf{x}|/|\mathbf{y}|$ is an infinitesimal.

For example, if ✶ is infinitesimal, then ✶² is also. To prove this all we need to point out is that for any positive real r,

$$\text{✶}^2 = \text{✶}\text{✶} < \text{✶}1 = \text{✶} < r,$$

since ✶ < 1. Let $\mathbf{x} = 2 + \text{✶}^2$ and $\mathbf{y} = 2 + \text{✶}$. Then **x** and **y** are infinitely close since $|\mathbf{x}-\mathbf{y}| = |\text{✶}-\text{✶}^2|$ is an infinitesimal. Both **x** and **y** are infinitely close to 2, but **x** is infinitely closer to 2 than **y** because $|\mathbf{x} - 2|/|\mathbf{y} - 2| = \text{✶}^2/\text{✶} = \text{✶}$ is infinitesimal. The following problems are more results of this type.

Problems

1) Prove that the square root of an infinitesimal is an infinitesimal.
   (*Hint*: Use the result that "if $\mathbf{x} < \mathbf{y}$ then $\sqrt{\mathbf{x}} < \sqrt{\mathbf{y}}$". First prove that this is true for hyperreals.)

2) Prove that an infinitesimal hyperreal times a finite hyperreal is an infinitesimal.

3) Prove that $\approx$ is an equivalence relation.

4) Show that a finite hyperreal **x** is surrounded by a cloud of hyperreals, infinitely close to it, that is, if **y** and **z** are infinitely close to **x**, then so is any hyperreal between **y** and **z**.

5) Let ✪ be an infinite hyperreal. Prove that ✪² is infinitely bigger than ✪.

6) Let ● be an infinite hyperreal. Find an infinite hyperreal that is infinitely bigger than all the powers ●^{n} for (finite) integral n.

7) If $f(x)$ is bounded for all real x, then is $f(\mathbf{x})$ bounded for all hyperreal $\mathbf{x}$?

   (*Hint*: $f(x)$ is a real function; $f(\mathbf{x})$ is the corresponding hyperreal function. A direct proof is possible using sequences of reals. Or use the transfer principle.)

8) If $f(x)$ is an integer for all real x, is $f(\mathbf{x})$ an integer for all hyperreal $\mathbf{x}$?

9) If $f(x)$ is finite for all real x, is $f(\mathbf{x})$ finite for all hyperreal $\mathbf{x}$?

10) Prove the rational hyperreals are dense in the hyperreals.

11) Express the Archimedean property of the reals in $\mathcal{L}$.

12) Since the Archimedean property is true for the reals, it follows from the previous exercise that it is also true for the hyperreals. This appears to contradict the fact that $\mathbb{H}$ contains infinitely large numbers. Resolve this paradox.

   (*Hint*: The Archimedean property says "Given two positive numbers a and b, there is a positive integer n such that $b < na$." The meaning of "there is an integer" shifts depending on whether the sentence is applied to $\mathbb{R}$ or to $\mathbb{H}$.)

Drawing the hyperreal line

Based on these problems, a picture of the hyperreal line like that in Figure 6.5.1 begins to emerge. In Figure 6.5.1, the reals, which are the same as the finite hyperreals (see Theorem 6.6.1) occupy the central portion of the hyperreal line. Each real is surrounded by a cloud consisting of the hyperreals infinitely close to it. The reals as a whole are surrounded by satellite lines,

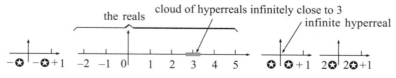

Figure 6.5.1. The hyperreal line.

6.5. The Nature of the Hyperreal Line

each centered about an infinitely large number and consisting of that number plus and minus the finite hyperreals. The satellite lines are each exactly as long as the real line itself.

Figure 6.5.1 uses tick marks at unit intervals along the line. Figure 6.5.2 gives a different view by using an infinite unit. On this scale the central part of the hyperreal line is occupied by an infinite hyperreal ⊘ and its finite real multiples. Satellite lines are still present, but based on infinite quantities, like $⊘^2$, which are infinitely larger than ⊘. At this scale the reals appear to be infinitely close to the origin. They are drawn as a cloud around 0.

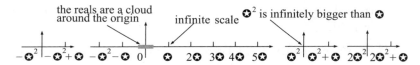

Figure 6.5.2. Zooming out on the hyperreals.

Figure 6.5.3 is $\mathbb{H}$ drawn with tick marks at an infinitesimal unit interval. On this scale, the hyperreals infinitely close to the origin are blown up to the size of the reals themselves (depicted in Figure 6.5.1). The satellite lines are now centered about other reals.

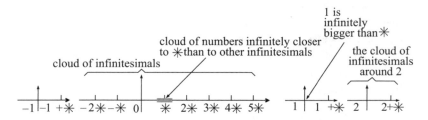

Figure 6.5.3. Zooming in on the hyperreals.

We must be careful because these figures leave out many features of the hyperreals, such as infinitesimals smaller than ✳ (like $✳^3$, and $\exp(-⊘)$), and infinite numbers larger than $⊘^2$ like $⊘^3$, $\exp(⊘)$, and even $⊘!$). There are also intermediate numbers. For example $⊘/2$ is larger than all reals r, but smaller than the hyperreals in the first satellite to the right of the reals in Figure 6.5.1, consisting of the hyperreals $⊘ \pm r$. Also, the clouds in the figures are grossly exaggerated. Each is really concentrated right at the single point it surrounds.

Although its internal structure is more complex than the reals, the hyperreal line and the real line are alike in one fundamental respect: both look the same at every scale.

Problems

12) Let $\otimes$ be infinitely large. Show that $\otimes/2 < \otimes - r$ for all reals r.

13) Let $\divideontimes$ be an infinitesimal. Draw a picture of $\mathbb{H}$ showing the relative location of these numbers: $\divideontimes$, $\divideontimes^2$, $\ln(\divideontimes)$, $\sqrt{\divideontimes}$, $\sin(\divideontimes)$, $\divideontimes^{\divideontimes}$, and $\exp(\divideontimes)$. (*Hint*: Using $\divideontimes = \{1, 1/2, 1/3, \ldots\}$ one can actually calculate these quantities.)

14) Let $\otimes$ be infinitely large. Draw a picture of $\mathbb{H}$ showing the relative locations of $\otimes$, $\otimes/2$, $\otimes^2$, $\ln(\otimes)$, $\exp(\otimes)$, $\sqrt{\otimes}$, $\otimes^{\otimes}$, $\sin(\otimes)$, and $\otimes!$.

15) Is the product of an infinitesimal and an infinitely large number ever finite? Is it always finite?

Decimal representation of hyperreals

Considering the strange numbers that $\mathbb{H}$ contains, it may seem surprising that hyperreals have decimal representations. These expansions portray the differences between the real and the hyperreal number systems from a different point of view, supplementing the drawings given above. Hyperreal decimals are also needed to define hyperreal continuous functions and the hyperreal calculus.

Theorem 6.5.1. *Every positive hyperreal* $\mathbf{x}$ *has a decimal representation*

$$\mathbf{x} = \mathbf{i}.d_1 d_2 d_3 \ldots d_\mathbf{n} \ldots,$$

where $\mathbf{i}$ *is a hyperreal integer and, for every positive hyperreal integer* $\mathbf{n}$, $d_\mathbf{n}$ *is a numeral* (0, 1, 2, 3, 4, 5, 6, 7, 8, *or* 9).

Proof. (By the transfer principle) This requires two functions:

$$i(x) = \text{ the integer part of the decimal expansion of } x,$$

and

$$d(n, x) = \text{ the } n\text{th fractional place in the expansion of } x.$$

6.5. The Nature of the Hyperreal Line

These are well-defined for real x, provided we forbid decimal expansions that end with repeated 9s. Together these two functions give the whole decimal expansion of the real x. Because $\mathbb{H}$ is a structure for $\mathfrak{L}$, both i and d extend to hyperreal $\mathbf{x}$ and hyperreal $\mathbf{n}$. The rest of the proof simply transfers properties of i and d from the structure $\mathbb{R}$ to the structure $\mathbb{H}$.

For example, let $\text{Int}(n)$ be the relation "n is an integer". Then

$$\forall x (\text{Int}(i(x)) \wedge (i(x) \leq x) \wedge (x - i(x) < 1))$$

says that $i(x)$ is the nearest integer less than or equal to x. Let $\text{Num}(n)$ be the relation "n is a numeral", i.e., "n is $0, 1, 2, \ldots,$ or 9". Then

$$\forall x \forall n (\text{Int}(n) \wedge (n > 0)) \rightarrow \text{Num}(d(n, x)))$$

says that for all reals x and integers n, $d(n, x)$ is a numeral. These statements are true for $\mathbb{R}$, hence true for $\mathbb{H}$. Finally, the statement

$$\forall x (|x - i(x) - d(1, x)/10| < 1/10)$$

says that $(i(x) + d(1, x)/10)$ approximates $\mathbf{x}$ to the nearest tenth. This characterizes $d(1, x)$ as the tenths place of the decimal expansion of $\mathbf{x}$. A similar statement can be made for each place. □

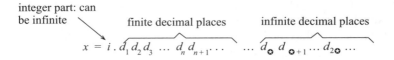

Figure 6.5.4. A hyperreal decimal expansion.

Problems

16) Find the hyperreal decimal expansion of $1/3$.

  (*Hint*: What is the usual decimal expansion of $1/3$? Express this in $\mathfrak{L}$.)

17) Find the hyperreal decimal expansion of $1/7$.

18) Describe the decimal expansion of an infinitesimal $*$.

19) Describe the decimal expansion of a hyperreal infinitely close to 3.

20) Describe the effect of multiplication by 10 on the decimal expansion of a hyperreal **x**. Describe the effect of multiplication by $10^✪$ where ✪ is an infinite integer.

21) Do hyperreals have decimal expansions to bases other than 10?

22) Do hyperreals have decimal expansions to an infinite base?

23) Prove that the hyperreals are dense in the reals, that is, between every two reals there is a non-real hyperreal.

6.6 The Hyperreal Calculus
Limits

The earliest use of infinitesimals was in the calculus. For centuries they were used to motivate results and set up applications. Here we show how infinitesimals are useful for proving theorems of the calculus.

We begin with limits of sequences. Let $\{a_n\}$ be a sequence of real numbers, which we may write out,

$$a_1, a_2, a_3, a_4, \ldots, a_n, \ldots .$$

Such a sequence is a real-valued function of the variable n. The variable n is not written between parentheses as is usual for functions. Nevertheless n is a variable for which we plug in different integer values. For a given n (25, say) the value of the function $\{a_n\}$ sits in the sequence at the nth place (i.e., it is a_{25}). As a function, the domain of $\{a_n\}$ is restricted to positive integers, still, like every real-valued function, $\{a_n\}$ has an extension to a hyperreal function. The hyperreal version is defined for all positive hyperreal integers, in particular, has values for infinite integers. As a hyperreal function $\{a_n\}$ looks like this:

$$a_1, a_2, a_3, a_4, \ldots, a_n, \ldots a_✪, a_{✪+1}, \ldots .$$

This process is exactly the way the decimal expansion of the reals generalized to the hyperreals. Compare the previous equation with Figure 6.5.4.

This leads to the hyperreal definition of limit.

Definition. A sequence $\{a_n\}$ **converges** to the limit b if $a_\mathbf{n}$ is infinitely close to b for all infinite integers $\mathbf{n} > 0$. A sequence $\{a_n\}$ is **Cauchy** if $a_\mathbf{n} \approx a_\mathbf{m}$ for all infinite integers $\mathbf{n}$ and $\mathbf{m} > 0$.

6.6. The Hyperreal Calculus

How neatly the hyperreal definitions capture the intuitive essence of limit and of Cauchy sequence! The classical notation for limit of a sequence,

$$b = \lim_{n \to \infty} a_n,$$

suggests that a limit is something that happens at infinity, but infinity is only a vague concept in real terms. In the hyperreals, however, there is a place called infinity (in fact, more than one) that can be plugged into formulas. The limit b of a sequence $\{a_n\}$ is $a_\otimes$, or at least is infinitely close to $a_\otimes$.

Problems

1) For the sequence $\{a_n\} = \{(n^2/(1+n)\}$ find $a_\otimes$.

2) Prove that the classical $\varepsilon - \delta$ definition of limit (see section 1.4) is equivalent to the hyperreal definition.

3) Prove that the classical definition of Cauchy sequence (given in section 1.4) is equivalent to the hyperreal definition.

4) Prove that a convergent sequence is Cauchy using hyperreal definitions.

Finding limits using hyperreals

Limits are simplified conceptually by the hyperreals but there are technical problems. Suppose we want to find the limit b of $\{a_n\}$. Writing $b = a_\otimes$ is useless in real terms because $a_\otimes$ is a hyperreal that may not be real, in fact, probably is not real. The next theorem provides a mechanism for computing the real limit of $\{a_n\}$.

Theorem 6.6.1. *Every finite hyperreal is infinitely close to a unique ordinary real r.*

Definition. Let **x** be a finite hyperreal, The unique ordinary real infinitely close to **x** is called the **real approximation** to **x** and written [**x**].

Real approximation is the hyperreal tool for the calculation of real limits. If $\{a_n\}$ is convergent, we write $b = [a_\otimes]$ for the limit. To prove that $\{a_n\}$ is convergent, we show that b is independent of the choice of the infinite integer $\otimes$. Using hyperreals, the trials and tribulations of the $\varepsilon - \delta$ definition of limit are replaced by the difficulty of computing real approximations. The problems explore some of these.

Problems

5) Let **x** be a hyperreal and let i and d_n be, respectively, the integer and fractional places in its decimal representation. If $i = 0$ and also $d_n = 0$ for all finite n, show that **x** is an infinitesimal.

   (*Hint*: If x is real, and $i = 0$ and $d_n = 0$ for the first K places, then $x < 1/10^K$. Express this in $\mathfrak{L}$ and apply it to a hyperreal.)

6) Prove that two ordinary reals are never infinitesimally close. Use this to prove that the real approximation to a finite hyperreal is unique.

7) Prove Theorem 6.6.1,

   (*Hint*: Let **x** be a hyperreal. Let i and d_n be the integer and fractional places of the decimal representation of **x**. Since **x** is finite, i is finite. If

   $$\mathbf{x} = i.d_1 d_2 d_3 \ldots d_{\bigodot} d_{\bigodot+1} \ldots d_{2\bigodot} \ldots,$$

   define

   $$r = i.d_1 d_2 d_3 \ldots,$$

   leaving off the infinite places of **x** in order to create a real number. Prove that r is an ordinary real and that the difference $\mathbf{x} - \mathbf{r}$ is an infinitesimal, where **r** is the hyperreal corresponding to the real r. This proves r is the real number we seek. Is r unique?)

8) Prove that if r is a real and $\ast$ is an infinitesimal, then $[r + \ast] = r$.

9) Prove that $[\mathbf{x}_1 + \mathbf{x}_2] = [\mathbf{x}_1] + [\mathbf{x}_2]$ and $[\mathbf{x}_1 \mathbf{x}_2] = [\mathbf{x}_1][\mathbf{x}_2]$. To what theorems from the usual theory of limits do these correspond?

10) Prove that if $\mathbf{x} \geq 0$, then $[\mathbf{x}] \geq 0$. Is it true that if $\mathbf{x} > 0$, then $[\mathbf{x}] > 0$?

    (*Note*: More generally, if **x** is in the open interval (a, b) then $[\mathbf{x}]$ is in the closed interval $[a, b]$.)

11) Prove that if $\mathbf{x} \neq 0$, then $[1/\mathbf{x}] = 1/[\mathbf{x}]$.

    (*Hint*: Let $r = [\mathbf{x}]$. Then $\mathbf{x} = r + \ast$, for some infinitesimal $\ast$. Investigate $1/(r + \ast) - 1/r$.)

6.6. The Hyperreal Calculus

Continuous functions

Intuitively a function $f(x)$ is continuous if a small change in x produces a small change in $y = f(x)$. Nothing can be smaller than an infinitesimal change. This suggests the following definition.

Definition. Let f be a real-valued function. Then f is **continuous at a real point** r if for every **x** infinitely close to r, $f(\mathbf{x})$ is infinitely close to $f(r)$. In symbols, if $\mathbf{x} \approx r$ then $f(\mathbf{x}) \approx f(r)$.

A real-valued function f is **continuous on an interval** if f is continuous at every real point in the interval.

The hyperreal definition of continuity says that an infinitesimal change in x produces an infinitesimal change in $f(x)$. This definition is easy to use. Here, for example, is a proof of the intermediate value theorem.

Theorem 6.6.2. *(The intermediate value theorem). Let a and b be reals. Let f be a continuous function defined on $[a, b]$ such that $f(a) < 0$ and $f(b) > 0$. Then there is a real number c between a and b such that $f(c) = 0$.*

Proof. Let n be a positive integer. If the interval from a to b is divided into n equal subintervals, the function f will change sign on one of them, as it does on the whole interval. Using real numbers, only finite subdivisions can be created. Using hyperreals, infinite subdivision is possible.

To do so we describe the subdivision process in $\mathfrak{L}$. First we write

$$\forall n((I(n) \wedge (n > 0)) \rightarrow \exists k(I(k) \wedge (0 \le k < n)$$
$$\wedge (f(a + k\frac{b-a}{n}) \le 0)$$
$$\wedge (f(a + (k+1)\frac{b-a}{n})) \ge 0))),$$

which says that for any positive integer n, if the interval $[a, b]$ is divided into n subintervals:

$$[a + k\frac{b-a}{n}, a + (k+1)\frac{b-a}{n}]$$

of length $(b-a)/n$, where $k = 0, 1, 2, \ldots, n-1$, then f changes sign on at least one of them.

By the transfer principle, this applies to the hyperreals and n can be chosen to be infinite, say $\mathbf{\Omega}$. It follows that there is an integer $\mathbf{k}$, $0 \le \mathbf{k} < \mathbf{\Omega}$,

such that
$$f(a + \mathbf{k}(b-a)/\Omega) \leq 0, \qquad (*)$$
and
$$f(a + (\mathbf{k}+1)(b-a)/\Omega)) \geq 0. \qquad (**)$$
Since $0 \leq \mathbf{k} < \Omega$,
$$a \leq a + \mathbf{k}\frac{b-a}{\Omega} < b,$$
and, if we let $c = [a + \mathbf{k}(b-a)/\Omega]$, then according to problem 7,
$$a \leq c \leq b.$$
Since $1/\Omega$ is an infinitesimal, c also equals $[a + (\mathbf{k}+1)(b-a)/\Omega]$. Now, using problem 12 and inequality (*),
$$f(c) = f([a + \mathbf{k}\frac{b-a}{\Omega}]) = [f(a + \mathbf{k}\frac{b-a}{\Omega})] \leq 0,$$
because f is continuous. Similarly, (**) implies that $f(c) \geq 0$. Thus, $f(c) = 0$. □

Problems

12) Let f and g be continuous functions. Use the hyperreal definition to prove that $f+g$, fg, and f composed with g are also continuous.

13) Prove that a real-valued function f is continuous if, and only if, $[f(\mathbf{x})] = f([\mathbf{x}])$, where $\mathbf{x}$ is a finite hyperreal in the domain of f.

14) Prove that a finite hyperreal integer is a real integer.

15) (*Boundedness of continuous functions*) Prove, using hyperreals, that every real-valued continuous function f defined on a closed interval $[a, b]$ is bounded above, that is there is a constant K such that $|f(x)| \leq K$ for all x in $[a, b]$.

(*Hint*: Given a continuous function defined on $[a, b]$, the goal is to prove the statement $\exists K \forall x((a \leq x \leq b) \rightarrow (-K < f(x) < K))$. Prove that it is true for $\mathbb{H}$ first. Remember that in $\mathbb{H}$, K can be infinite, but because f is a continuous function, $f(x)$ will be finite for $a \leq x \leq b$.)

16) (*Maximum value theorem*) Prove, using hyperreals, that, if f is a continuous function on $[a, b]$, then there is a point c in this interval such that $f(x) \leq f(c)$ for all x in $[a, b]$.

(*Hint*: Use a subdivision argument.)

6.6. The Hyperreal Calculus

Hyperreal Differentiation

Using hyperreals, the derivative can be defined as a quotient of infinitesimals, as envisioned by Leibniz:

Definition. Let f be a real-valued function defined on an open interval (a, b). If dx is an infinitesimal and x is in (a, b), we write $dy = f(x + dx) - f(x)$ for the change in the value of f at x. The **difference quotient** is the ratio

$$f'(x) = \left[\frac{dy}{dx}\right] = \left[\frac{f(x + dx) - f(x)}{dx}\right].$$

If this real quantity is independent of dx, then we say that f is **differentiable at** x and $f'(x)$ is called the **derivative** of f.

This definition captures the essence of the derivative as a rate of change and makes it fairly easy to compute derivatives. For example, here is the derivative of $1/x$:

$$\left(\frac{1}{x}\right)' = \left[\frac{\frac{1}{x+dx} - \frac{1}{x}}{dx}\right] = \left[\frac{1}{dx}\frac{x - (x + dx)}{x(x + dx)}\right]$$

$$= \left[\frac{1}{dx}\frac{-dx}{x(x + dx)}\right] = \left[\frac{-1}{x(x + dx)}\right]$$

$$= \frac{-1}{[x][x + dx]} = \frac{-1}{x^2},$$

a straightforward calculation with hyperreals. In $\mathbb{H}$, we may divide by dx because it is not zero. Later we cancel it. Since the result of this computation is independent of dx, we conclude that $1/x$ *is* differentiable with derivative $1/x^2$. A limit is replaced by a real approximation.

Problems

17) Differentiate these functions using hyperreals.

   (a) x^3
   (b) $(x + 1)/x^3$

18) Prove that a function differentiable at a real number x is continuous at x.

19) Prove the sum and product rules for differentiation:

(a) $(f(x) + g(x))' = f'(x) + g'(x)$
(b) $(f(x)g(x))' = f(x)g'(x) + f'(x)g(x)$

20) Prove the chain rule: $f(g(x))' = f'(g(x))g'(x)$.

21) Prove that $[\sin(\ast)/\ast] = 1$ where $\ast$ is an infinitesimal.

(*Hint*: If the real number r is less than $\pi/2$, then $\sin(r) \leq r \leq \tan(r)$. Transfer to $\mathbb{H}$.)

21) Find the derivative of $\sin(x)$.

(*Hint*: Use the addition theorem:

$$\sin(x + y) = \sin(x)\cos(y) + \sin(y)\cos(x).)$$

6.7 Construction of an Ultrafilter

Well-Ordered sets

We complete the construction of the hyperreals by proving the existence of an ultrafilter. The proof uses a powerful, proof technique called transcendental induction, which requires knowledge of well-ordered sets. We begin with the definition:

Definition. A set S is **partially ordered** if it has a relation $\leq$ satisfying

(a) For every x in S, $x \leq x$, — **reflexivity**
(b) For x, y in S, if $x \leq y$ and $y \leq x$, then $x = y$, — **anti-symmetry**
(c) For x, y, z in S, if $x \leq y$ and $y \leq z$, then $x \leq z$, — **transitivity**

A partial order is weaker than a linear order. The reflexive and anti-symmetric laws together are not enough to imply trichotomy.

A new feature of partially ordered sets is the possibility of incomparable elements. Two elements x and y are **comparable** if either $x \leq y$ or $y \leq x$. In a linearly ordered set, every pair of elements is incomparable, but a typical partially ordered set has a branching, tree-like structure. Elements in different branches are incomparable. Figure 6.7.1 gives an example. Here $b > d$, but b and c are incomparable.

Definition. A subset X of a partially ordered set has a **least element** c if c is in X and $c \leq x$ for all x in X. A partially ordered set is **well-ordered** if every non-empty subset of X has a least element.

6.7. Construction of an Ultrafilter

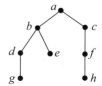

Figure 6.7.1. A partially ordered set.

As an example of a well-ordered set, take any finite set, for example the letters of the alphabet, and put them in linear order:

$$a, b, c, d, e, f, g, h, i, j, k, l, m, n, o, p, q, r, s, t, u, v, w, x, y, z.$$

Given a subset X of S, if we read across the list of the elements of S, the first element of X we encounter is the least element. Therefore, S is well-ordered.

For finite sets, 'linear ordered' and 'well-ordered' are equivalent, but not for infinite sets. The integers, for example, are linearly ordered, but not well-ordered. The reals are another set that is linearly ordered, but not well-ordered.

Problems

1) Prove that if x and y are comparable elements from a partially ordered set, then exactly one of the following is true: $x < y$, or $y < x$, or $x = y$. Thus the trichotomy law holds for comparable elements.

2) Let X be a subset of a partially ordered set S. Suppose that X has a least element. Prove that it is unique.

3) Prove that the natural numbers $\mathbb{N}$ are well-ordered.

4) Show that the integers $\mathbb{Z}$ are not well-ordered.

5) Show that the unit interval $[0, 1]$ is not well-ordered.

Infinite well-ordered sets

The natural numbers $\mathbb{N}$ are an infinite well-ordered set. A more exotic example is obtained by adding an extra element, say ω. Call the new set W. It is well-ordered by using the usual order for the natural numbers and stipulating that $\omega > x$ for all x in $\mathbb{N}$. The set W therefore looks something like:

$$W : 1, 2, 3, 4, \ldots, \ldots, \ldots, \omega.$$

In W, ω has no immediate predecessor (like 1). More examples like W are concocted in the next problems.

Problems

6) Let S be a well-ordered set. Show that S is linearly ordered, in other words a well-ordered set satisfies the trichotomy law.

7) Let S be a well-ordered set. An element x of S is **maximal** if there is no element y in S such that $x < y$. Show that a well-ordered set S can have at most one maximal element, but need not have any.

8) Let S be a well-ordered set. An element x of S has a **successor** if there is an element y in S such that $x < y$ and there is no element z such that $x < z < y$. Show that every non-maximal element in S has a successor.

9) Prove that the set W (above) is well-ordered.

10) Find an example of a well-ordered set S in which four elements are without an immediate predecessor.

11) Describe an example of a well-ordered set S in which an infinite number of elements are without an immediate predecessor.

Transfinite induction and ordinary induction

Every well-ordered set is the basis for a kind of proof by induction, called transfinite induction.

Theorem 6.7.1. *(Principle of transfinite induction). Let S be a well-ordered set. Let $s(x)$ be a statement about an element x of S. Suppose that s has the* **inductive property**:

if $s(b)$ is true for all $b < a$, then $s(a)$ is true.

Then $s(x)$ is true for all x in S.

Proof. Let X be the set of all x in S for which $s(x)$ is false. Suppose, contrary to what we want to prove, that X is non-empty. Since S is well-ordered, X has a least element, say a. But the definition of a means that $s(b)$ is true for all $b < a$. By the inductive property, $s(a)$ is true, contradicting the assumption that $s(a)$ is false. This contradiction proves the theorem. □

6.7. Construction of an Ultrafilter

Ordinary induction is transfinite induction applied to the natural numbers. Assuming that the natural numbers are well-ordered, Theorem 6.7.1 implies mathematical induction. Transfinite induction replaces the two cases of ordinary induction by a single case which applies to an element of a well-ordered set whether or not it has an immediate predecessor.

Corollary. *Let $s(n)$ be a statement about an element of the set $\mathbb{N}$. If $s(n)$ has the following two properties:*

(a) Base Case: $s(1)$ is true,

(b) Inductive Case: if $s(n)$ is true then $s(n+1)$ is true,

then $s(n)$ is true for all elements of $\mathbb{N}$.

Proof by transfinite induction

Transfinite induction is a genuinely infinite form of induction. Ordinary mathematical induction does prove an infinite number of statements, one for each natural number. But each of those proofs involves only a finite number of steps. Transfinite induction, on the other hand, can prove families of statements in which it takes an infinite number of steps to reach some of the statements. For example, to reach the statement $s(\omega)$, where ω is the last element in the well-ordered set W defined above, it is necessary to pass through the infinite list of statements: $s(1), s(2), s(3) \ldots$ coming before it.

The well-ordering principle

To construct the hyperreals, we make this assumption:

The Well-Ordering Principle. *Every set can be well-ordered.*

Informally, one can defend the well-ordering principle as follows. Given a set S, obviously one can choose any element, say x_1, to be the least. After removing x_1 from S, another element, x_2, can be chosen to be x_1's successor. Continuing in this way, removing elements and choosing, eventually S becomes well-ordered.

For finite sets this is fine, but for infinite sets there is no basis for making the infinite number of choices needed to complete the process without further set theory assumptions. This doesn't prove the well-ordering principle, but only makes it more plausible—perhaps.

The well-ordering principle plays an important role in classical mathematics, particularly in abstract algebra and analysis. It tends to be used in several equivalent forms, most notably the axiom of choice and Zorn's lemma. In any form, it is controversial. Regarded by most mathematicians as an important mathematical principle, it tends to be used sparingly. It cannot be deduced from the usual axioms of set theory. We must assume it to prove certain theorems. The existence of an ultrafilter is an example.

To a constructivist (see Chapter 5) the well-ordering principle is unacceptable. It assumes into existence something that can never be constructed. In this chapter we regard it as a fundamental assumption of mathematics to be used as needed.

Problems

12) Prove the corollary.

    (*Hint*: Two proofs are possible. A direct proof can be based on the fact that $\mathbb{N}$ is well-ordered, imitating the proof of the principle of transcendental induction. Ordinary induction can also be deduced from transcendental induction by showing that the inductive property is equivalent, for $\mathbb{N}$, to the two conditions of ordinary induction.)

13) Well-order the integers.

14) Prove that every countably infinite set can be well-ordered.

    (*Hint*: What is the definition of countably infinite?)

15) The reals are not well-ordered by their usual order. To well-order the reals requires an unusual order relation. Describe, as best you can, what this order might be like.

Ultrafilters exist!

Recall that an ultrafilter is a collection $\mathfrak{U}$ of subsets of $\mathbb{N}$ with the four properties:

(a) $\mathfrak{U}$ contains the cofinite sets, but no finite set,

(b) if A and B are in $\mathfrak{U}$, then $A \cap B$ is in $\mathfrak{U}$,

(c) if A is in $\mathfrak{U}$ and B contains A, then B is in $\mathfrak{U}$,

(d) for every subset A of $\mathbb{N}$, either A or the complement of A is in $\mathfrak{U}$.

6.7. Construction of an Ultrafilter

It will be useful to have a name for a collection $\mathfrak{P}$ of subsets of $\mathbb{N}$ that is not an ultrafilter, but satisfies the first two properties:

(a) $\mathfrak{P}$ contains the cofinite sets, but no finite set,

(b) if A and B are in $\mathfrak{P}$, then $A \cap B$ is in $\mathfrak{P}$.

Such a collection is called a **prefilter**. Prefilters exist; the cofinite sets are an example.

Lemma 1 *If a prefilter $\mathfrak{P}$ satisfies (d), then $\mathfrak{P}$ is an ultrafilter.*

Lemma 1 says that the hard part of constructing an ultrafilter is satisfying (d). The next lemma shows how to start with a prefilter and take one step forwards.

Lemma 2 *Let $\mathfrak{P}$ be a prefilter and let A be a subset of $\mathbb{N}$ such that neither A nor the complement of A is in $\mathfrak{P}$. Then there exists a prefilter $\mathfrak{P}^+$ containing the sets of $\mathfrak{P}$ plus either A or the complement of A.*

Theorem 6.7.2. *There exists an ultrafilter.*

Proof. (By transfinite induction) The essence of the proof is showing that it is possible to take an infinite number of steps (such as the one described in Lemma 2) starting with a prefilter, such as the cofinite sets, adding additional sets one at a time, ending with an ultrafilter.

Let $\mathfrak{S}$ be the set of *all* subsets of $\mathbb{N}$. Well-order $\mathfrak{S}$. We have just well-ordered an uncountably infinite set.

Let the well-ordering be represented by the symbol $\implies$. We will construct, for each set A in S, a prefilter $\mathfrak{P}(A)$ such that (a) $\mathfrak{P}(A)$ contains A or the complement of A, and (b) if $B \implies A$, then $\mathfrak{P}(B)$ is contained in $\mathfrak{P}(A)$. The proof is by transfinite induction on S.

Let $s(A)$ be the statement "The prefilter $\mathfrak{P}(A)$ exists." To prove $s(A)$ we must construct $\mathfrak{P}(A)$, but we get to assume that $\mathfrak{P}(B)$ exists for $B \implies A$. There are two cases. If A has an immediate predecessor, then Lemma 2 shows how to extend $\mathfrak{P}(B)$ to $\mathfrak{P}(A)$ in one step. Otherwise, A has no immediate predecessor. In this case, let $\mathfrak{P}$ be the union of all $\mathfrak{P}(B)$ for $B \implies A$. Then $\mathfrak{P}$ is a prefilter (according to problem 18) and applying the process of Lemma 2 gives us $\mathfrak{P}(A)$.

Transfinite induction now implies that $s(A)$ is true for all subsets A of $\mathbb{N}$. If we let $\mathfrak{U}$ be the union of the prefilters $\mathfrak{P}(A)$ for all subsets A of $\mathbb{N}$, then $\mathfrak{U}$ is a prefilter (see problem 18) satisfying property (d). Lemma 1 implies $\mathfrak{U}$ is an ultrafilter. $\square$

Problems

16) Prove Lemma 1.

    (*Hint*: Try proof by contradiction.)

17) Prove Lemma 2.

    (*Hint*: Show that either A or A complement has non-empty intersection with all the sets in $\mathfrak{P}$. Having established this, suppose, for simplicity, that this set is A. Then set

    $$\mathfrak{P}^+ = \mathfrak{P} \cup \{A \cap B | B \in \mathfrak{P}\}$$

    and prove that $\mathfrak{P}^+$ is a prefilter.)

18) Prove for any A in S, that the union of all $\mathfrak{P}(B)$, where $B \implies A$, is a prefilter. Complete the proof that an ultrafilter exists by proving that the union of all the prefilters $\mathfrak{P}(A)$ is a prefilter.

19) Prove that every vector space has a basis.

    (*Hint*: Well-order the elements of the vector space and use transfinite induction.)

Do we really need an ultrafilter?

An ultrafilter is a peculiar mathematical object, and the principles upon which its existence is based, transfinite induction and well-ordering, are powerful tools, capable of proving strange results. To emphasize this point, here is a paradoxical theorem that can be proven only using transfinite induction.

Theorem 6.7.3. (*The Banach-Tarski paradox*). *Let S be a solid three-dimensional sphere. Then S can be partitioned into five sets T_1, T_2, T_3, T_4, and T_5 in such a way that T_1 and T_2 can be assembled into a solid sphere S_1. Likewise sets T_3 and T_4 can be assembled into a solid sphere S_2. Both spheres S_1 and S_2 are identical to the original sphere S.*

At first this result is difficult to grasp. It appears that it is possible to create something out of nothing. Starting with one sphere, two spheres equal to the first are constructed by cutting and pasting alone. If this construction actually could be carried out with real spheres, think of the benefits for humanity! As one might suspect, the construction whose existence is asserted

6.7. Construction of an Ultrafilter

by the Banach-Tarski paradox cannot be accomplished physically. For one thing, the pieces T_1, T_2, and so forth, are too complicated to cut out of the sphere. Even if the physical act of cutting were possible, one couldn't even begin because the instructions how to do it are not known. No one knows the dimensions of the pieces, or what points of the sphere S go into which piece. The proof of existence of the pieces, like the construction of an ultrafilter, uses transfinite induction and the well-ordering principle, and gives no way to find the pieces it claims exist.

The Banach-Tarski paradox is paradoxical because it contradicts intuitive ideas about volume and solids. There is no contradiction, however. The usual mathematical concepts of size and volume don't apply to sets of the complexity created when proving Theorem 6.7.3.

To prove the existence of the hyperreals, we used an ultrafilter. We might hope that the hyperreals could be constructed and their properties established without using such complicated and controversial means. However, the next theorem shows that the existence of an ultrafilter is essential to the existence of the hyperreals.

Theorem 6.7.4. *A hyperreal number system, $\mathbb{H}$, satisfying the transfer principle exists if and only if there is an ultrafilter.*

Proof. We already know that the existence of an ultrafilter implies the existence of a hyperreal number system satisfying the transfer principle. Conversely, assume that $\mathbb{H}$ exists. Using results about real integers transferred to $\mathbb{H}$, it follows (by problem 20) that $\mathbb{H}$ contains an infinite natural number $\mathbf{\circleddash}$. Let A be a subset of the natural numbers, and let $R_A(n)$ be the relation that is true if and only if n is in A. By definition $\mathbb{H}$ is a structure for $\mathcal{L}$. Therefore R_A applies to $\mathbb{H}$. If we let

$$\mathfrak{U} = \{A | A \text{ is a subset of } \mathbb{N} \text{ and } R_A(\mathbf{\circleddash}) \text{ is true}\},$$

then $\mathfrak{U}$ is an ultrafilter (see problem 21). In other words, for every subset A of $\mathbb{N}$, there is a version of A in $\mathbb{H}$, and those A whose $\mathbb{H}$-versions contain $\mathbf{\circleddash}$ constitute an ultrafilter. □

Problems

20) Using the transfer principle alone, prove that a hyperreal number system satisfying the transfer principle contains an infinite natural number.

21) Complete the proof of Theorem 6.7.4 by proving that $\mathfrak{U}$ is an ultrafilter.

22) Show that the cofinite sets satisfy all the properties required of an ultrafilter but one.

23) We have gone to all this trouble (learning about well-ordered sets in order to use transfinite induction in order to construct an ultrafilter in order to construct the hyperreals) for the sake of only one measly property! What works and what doesn't work if we try to construct the hyperreals using the cofinite sets and forget about an ultrafilter?

6.8 A Final Word about the Hyperreals

The hyperreals embody the eternal fascination of the infinite. They also justify the use of infinitely large and infinitely small numbers for mathematical computation. Since their discovery, around 1961, by Abraham Robinson (1918–1974), the hyperreals have been the subject of research and employed as a tool in other branches of mathematics.

The impact on the calculus and elsewhere of this newly legitimized use of infinitesimals has been far from revolutionary, however. Due to their close connection with the reals (via the transfer principle), anything proven about the reals using hyperreals can also be proven without them. As far as the theory of the reals alone is concerned, nothing new arises through use of hyperreals. On one hand, this is reassuring; it confirms the soundness of centuries of intuitive mathematical work with infinitesimals. On the other hand, it does tend to make the development of the hyperreal calculus less exciting. New results have been discovered using hyperreals, nevertheless, and their proofs subsequently translated into more conventional terms.

References for the hyperreals: This chapter is based on [I2]. For more depth, see [I3], which also has an excellent annotated bibliography.

7

The Surreals

Introduction

The surreals, like the hyperreals, are a relatively recent invention, discovered in the early 1970s by John H. Conway. Like the hyperreals, the surreals contain infinitesimally small numbers and infinitely large numbers. Unlike the hyperreals, the surreals are not a structure for the reals. Thus it is not true that every function and relation on the reals has an extension to or a meaning for the surreals.

A spirit of playfulness animates the surreals. They might be called hippie numbers after the "flower" children, who dropped out of society and lived communally in the Vietnam War period when the surreals were discovered.

A distinctive feature of the surreals is that they are part of the theory of combinatorial games. Within the world of combinatorial games, surreal numbers constitute a reference set of games used to evaluate positions in other games. The connection between the surreals and games leads to a proof technique unique to the surreals: prove a theorem by playing a game. Surreal proofs are constructive in spirit, although dependent on a strong form of induction.

Another feature of the surreals is that while constructing them, all other numbers are constructed too. The number systems we've considered so far have taken the rationals for granted. The surreals start with nothing. Not even the integers are needed.

7.1 Combinatorial Games

Combinatorial games are games between two opponents, traditionally called Left and Right, who take turns changing the position of the game. Some of these games are played on a board with physical pieces of some sort; others are purely mental. It is a central insight of the theory that to describe the position of a game all that is necessary is to describe what positions can result from a move by Left (called the **left options** of the position), and what positions can result from a move by Right (called the **right options**). In other words, the mathematical nature of a position X is completely determined by the lists of its left and right options, which themselves are positions.

If X is a position, we write

$$X = \{\ldots X^L \ldots | \ldots X^R \ldots\},$$

so that X^L stands for a typical left option and X^R stands for a typical right option. For example, if X is a position from which Left can move to A, B or C, while Right can move to D and E, we have

$$X = \{A, B, C | D, E\}.$$

What actually happens in this position depends on whose move it is.

A game ends when the player whose turn it is cannot play. That player is declared the loser; the other is the winner. At this point the position is either

$$X = \{X^L |\ \} \qquad \text{Right's turn and Left wins,}$$

or

$$X = \{\ |X^R\} \qquad \text{Left's turn and Right wins,}$$

and either the set of left options is empty or the set of right options is empty. As we learn later, the positions that take these forms are examples of surreal numbers, so we can use the number of the final position to evaluate the game. When Left wins, the value is positive or zero; when Right wins, the value is negative or zero.

To make proofs work, it is necessary to assume that in no game is there an infinite sequence of positions each of which is an option of its predecessor. In particular, as follows immediately from this assumption, no combinatorial game can go on forever.

Every game has an **initial position**, the position from which it is agreed that play starts. This position contains complete information on the whole game, since its options are the permitted second positions in the game,

7.1. Combinatorial Games

whose options in turn are the possible third positions in the game, and so on. Therefore, we identify each game with its initial position. Conversely, every position defines a game, the game played by starting from that position. Thus, the terms 'game' and 'position' are interchangeable. This discussion is summarized as follows:

Definition. If L and R are two lists of games, then $X = \{L|R\}$ is a **combinatorial game**, provided that X contains no infinite sequence of games each of which is an option of its predecessor. The games L are the **left options** of X and the games R are the **right options** of X.

The first games

The definition of combinatorial game is recursive. A game requires lists of the left and right options which must be games already defined. How does this start? What's the base case? Either L or R (or both) can be the empty list of games. Thus the first game to be defined is the game **zero**:

$$0 = \{ \ | \ \},$$

with no move for Left or for Right.

We regard the game 0 as defined on day 0. On the next day, day 1, the sets L and R of left and right options can either be empty or contain 0, and so on day 1, three games are defined:

$$1 = \{0| \ \}, -1 = \{ \ |0\}, \text{ and } * = \{0|0\}.$$

As we see later, 1 and -1 are numbers, but $*$ is not.

Concerning the play of these games, it is essential to quote Conway [J4]:

> The simplest game of all is the *Endgame* 0. I courteously offer you the first move in this game, and call upon you to make it. You lose, of course, because 0 is defined as the game in which it is never legal to make a move.
>
> In the game $1 = \{0| \ \}$, there is a legal move for Left, which ends the game, but at no time is there any legal move for Right. If I play Left, and you Right, and you have first move again (only fair, as you lost the previous game) you will lose again, being unable to move even from the initial position. To demonstrate my skill, I shall now start from the same position, make my move to 0, and call upon you to make yours.
>
> Of course you are beginning to suspect that Left always wins, so for our next game, -1, you may play as Left and I as Right! For the last of

our examples, the new game $* = \{0|0\}$, you may play whichever role you wish, provided that for this privilege you allow me to play first.

The reader will have realized that:

—in the game 0, there is a winning strategy for the second player,

—in the game 1, there is a winning strategy for Left (whoever starts),

—in the game -1, there is a winning strategy for Right (ditto), and

—in the game $*$, there is a winning strategy for the first player.

It turns out that every combinatorial game fits into one of these four categories.

Meanwhile, on day 2 the following games (among others) are defined:

$$2 = \{0, 1|\ \}, -2 = \{\ |0, -1\}, 1/2 = \{0|1\},$$
$$\uparrow = \{0|*\}, \downarrow = \{*|0\}.$$

In general, every game is defined on a certain day and has for left and right options older games, that is, games created on earlier days.

Proof by Infinite Descent

The assumption that no game contains an infinite sequence of games of which each is an option of its predecessor justifies a kind of proof by induction called **proof by infinite descent**. Thus let $s(X)$ be some statement that we wish to prove is true of all games X. Then all we need prove is that

"if $s(Y)$ holds for all options Y of X,
then $s(X)$ holds."

If this is established, then $s(X)$ must hold for all games. For if $s(X)$ fails for a game X_0, then it must fail for some option X_1 of X_0 and then for some option X_2 of X_1, and so on. This leads to an infinite sequence of games each of which is an option of its predecessor. Since this is impossible, $s(X)$ holds for all games.

7.1. Combinatorial Games

Problems

1) Let us say a set is **nice** if all its proper subsets are nice. What sets are nice?

   (*Hint*: Is the empty set nice?)

2) Let us define a game as **easy** if all its options are easy. What games are easy?

3) Justify ordinary mathematical induction by using infinite descent.

   (*Hint*: Use proof by contradiction.)

4) Can infinite descent be used to justify transfinite induction?

Examples of games

Combinatorial game are pure strategy games of finite length. Many common games are excluded because moves in a combinatorial game cannot be made by chance (no dice or other random number generator is used), there are no ties, and the game cannot last forever. Chess is out (ties are possible), as well as most popular board games (dice are used). There are many interesting and difficult games, however, to which Conway's theory applies. For example, while overall the game of Go does not conform to Conway's theory (the winner is determined by a scoring system rather than whoever moves last), Go strategy can be analyzed (see [J3]).

Here are a few sample games to which this theory applies directly and which, in addition, have the advantage of being easily played with only pencil and paper.

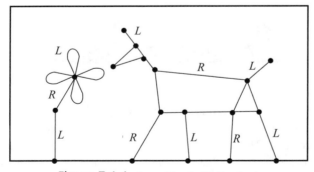

Figure 7.1.1. A position in Hackenbush.

Hackenbush. A Hackenbush position consists of a ground line or frame attached to which are a number of line figures. An example is in Figure 7.1.1.

A move by either Left or Right consists of removing (or hacking) a line. When a line is hacked, other lines which are thereby disconnected from the ground are also removed. Lines labeled L may only be hacked by Left; lines labeled R may only be hacked by Right; unlabeled lines may be hacked by either. Last player to move wins.

Dominos (or **CrossScram**). A position in this game is a portion of a chess or checker board. An example is in Figure 7.1.2. A move consists of the placement of a domino on the board. Left places dominos horizontally; Right vertically.

Figure 7.1.2. A position in Dominos.

Nim. A position consists of some piles of coins (see Figure 7.1.3). To move, a player removes any number of coins from one pile. This game is historically important as it was one of the first games to be given a complete theory.

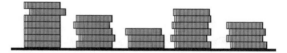

Figure 7.1.3. A position in Nim.

Col. A position in Col is a graph. A **graph** is a set of dots (called **vertexes**) some of which are connected by lines (called **edges**). An example is in Figure 7.1.4. The players take turns circling a vertex, each using a different mark. The rule is that adjacent (i.e., connected) vertexes may not be marked by the same player.

7.1. Combinatorial Games

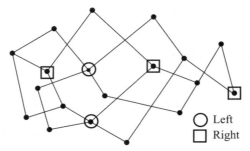

Figure 7.1.4. A Col position.

Snort. Snort is played like Col, but uses the rule that two adjacent vertexes may *only* be marked by the same player.

Problems

(*Suggestion*: The first time through this book, choose two games to study and ignore the others. One should be Hackenbush.)

5) Which of the games: $0, 1, -1, 2, -2, 1/2, -1/2, *, \uparrow$ and $\downarrow$ are realizable as Hackenbush positions?

6) Which of the games listed in problem 5 can be Domino positions?

7) Which of the games listed in problem 5 can be Nim positions?

8) Which of the games listed in problem 5 can be Col positions?

9) Which of the games listed in problem 5 can be Snort positions?

Summary

Combinatorial games are games of position in which each player, Left and Right, has a set of options or moves:

$$X = \{\ldots X^L \ldots | \ldots X^R \ldots\}.$$

Combinatorial games are created day by day starting only with the empty set. They are games of pure strategy that cannot go on forever.

7.2 The Preferential Ordering of Games

Conway created an order relation for games that expresses the concept of preference or advantage. If $X \geq Y$ in this order, then the game X offers some benefit or advantage for Left over playing game Y. Naturally Right then prefers Y to X. The problem is to express this in terms of left and right's options in X and in Y. Here is Conway's definition:

Definition. If X and Y are games, we say that **Left prefers** X **to** Y (or **Right prefers** Y **to** X) and write $X \geq Y$ if it never happens that

$$Y^L \geq X \text{ or } Y \geq X^R,$$

where Y^L represents any left option of Y and X^R represents any right option of X.

The order of X and Y is thus determined recursively by comparing X with Y's options and Y with X's options. This recursion is governed by two principles. Imagine the games being created day by day. As each new game X is created, its order (in comparison with a game Y already created) is determined in part by comparisons already established on earlier days between the already created options X^R of the new game and the earlier created game Y. This is the question whether $Y \geq X^R$. Eventually, all questions of order (including, whether $Y^L \geq X$) are settled by arguments using the empty set since if you play any game long enough either the set of left options or the set of right options becomes empty. Therefore it is important to realize that any statement about the empty set is true if it makes no assumption that the empty set contains anything. In particular, all statements using only universal quantification are true of the empty set.

Tactically, $X \geq Y$ says two natural things about X and Y. First, even if Left is given the first move in Y (therefore playing in some Y^L), Left prefers to play in X. Secondly, Left will not prefer Y even if forced to give Right the first move in X and therefore have to play in some X^R.

Operationally, the definition of order turns the single inequality $X \geq Y$ into the negation of two inequalities in which the roles of X and Y are switched and each in turn is replaced by an option. This observation is useful while mechanically checking order relations.

We also define

$X \leq Y$ if $Y \geq X$,   — X is **preferred** to Y **by Right**

$X = Y$ if $X \geq Y$ and $X \leq Y$,   — X **equals** Y

7.2. The Preferential Ordering of Games

and

$X || Y$ if neither $X \geq Y$ nor $X \leq Y$. — X and Y are **confused**

Equality is therefore a defined concept for games. We prove shortly that it is an equivalence relation. As with other situations where equality is defined, we must distinguish between equality and identity. Two games, X and Y, are identical (written $X \equiv Y$) if they have exactly the same left and right options. Equal games need not be identical, although, as we show eventually, equal games always have the same outcome and the same value.

The relation $||$ is needed because the ordering of games does not satisfy the trichotomy law. This is not a linear order. The relation $X || Y$ indicates that X and Y are **incomparable** or **confused** by the order.

Problems

---

1) Establish these relations:

   (a) $0 \geq 0$
   (b) $1 \geq 0$
   (c) $0 \geq -1$
   (d) $2 \geq 1 \geq 1/2 \geq 0$

   (*Example*: $1 \geq -1$. By definition this is true if Left prefers no left option of -1 to 1 and never prefers -1 to any right option of 1. We check first that not $-1^L \geq 1$. This is true because -1 has no left options. Similarly not $-1 > 1^R$ because 1 has no right options.)

2) Establish that the following relations are false:

   (a) $0 \geq 1$
   (b) $0 \geq 1/2$
   (c) $1/2 \geq 1$
   (d) $0 \geq *$
   (e) $* \geq 0$

   (*Example*: not $-1 \geq 1$. By definition this is true if either Left prefers a left option of 1 to -1 or prefers 1 to a right option of -1. In this case both are true. 0 is a left option of 1 and according to Problem 1(a), $0 \geq -1$. Similarly, 0 is a right option of -1 and $1 \geq 0$ by Problem 1(b).)

3) Establish these equalities:
   (a) $1 = \{-1, 0 | \ \}$
   (b) $2 = \{ 1 | \ \} = \{-1, 1 | \ \} = \{-1, 0, 1 | \ \}$
4) Establish these relations:
   (a) $1 \geq *$
   (b) $-1 \geq *$
   (c) $* || 0$

Equality of games is an equivalence relation

This follows immediately if we prove that the ordering of games is transitive:

Theorem 7.2.1. *The order relation $\leq$ is transitive.*

Proof. (By infinite descent) Suppose that $X \geq Y$ and $Y \geq Z$. This means that it is never true that

$$Y^L \geq X \text{ or } Y \geq X^R \text{ or } Z^L \geq Y \text{ or } Z \geq Y^R.$$

Our goal is to prove that $X \geq Z$. That is, we want to show that it is never true that either

$$Z^L \geq X \text{ or } Z \geq X^R.$$

We first prove the first. That Z^L appears here suggests that we use induction on Z. This means that we assume that the result is true for any option of Z. Thus we assume that $\geq$ is transitive for the options of Z, in particular for Z^L.

On top of this we use proof by contradiction. Suppose, contrary to what we want to prove, that it is true that

$$Z^L \geq X,$$

for some left option Z^L of Z. Combining this with the given fact that $X \geq Y$ and using the induction assumption, we conclude that

$$Z^L \geq Y.$$

This contradicts the given fact that $Z^L \geq Y$ never happens. Therefore the assumption $Z^L \geq X$ must be false. A similar proof (using induction on X) establishes that it is never true that $Z \geq X^R$. □

7.2. The Preferential Ordering of Games

Problems

5) Complete the proof of Theorem 7.2.1 by proving inductively that it is never true that $Z \geq X^R$.

6) For all games X, prove that

    (a) it is never true that $X \geq X^R$,

    (b) it is never true that $X^L \geq X$,

    (c) $X \geq X$.

   (*Hint*: Prove all three statements simultaneously by infinite descent.)

   (*Example*: Proof of (a). Assume that (a) – (c) are all true for the options of X. Now (a) says not $(Y \geq X^R)$, which by definition means that either $X^{RL} \geq X$ or $X^R \geq X^R$. By induction assumption (c), the second of these is true for the option X^R of X. Thus (c) for an option of X proves (a) for X itself.

7) Prove that $X = X$ for all games.

8) Prove that equality of games is an equivalence relation.

Order predicts the winner of a game

The next theorem connects order with game playing strategy.

Theorem 7.2.2. (*The classification theorem*). Let X be any game. Then

(a) $X > 0$ (X is **positive**, meaning $X \geq 0$ but X does not equal 0) if and only if there is a winning strategy for Left,

(b) $X < 0$ (X is **negative**, meaning $X \leq 0$ but X does not equal 0) if and only if there is a winning strategy for Right,

(c) $X = 0$ (X equals the zero game) if and only if there is a winning strategy for the second player, and

(d) $X || 0$ (X is **fuzzy**) if and only if there is a winning strategy for the first player.

Proof. (By simultaneous infinite descent) We illustrate by proving (a).

Proof of (a): Let X be a positive game. This means that $X \geq 0$ and not $(0 \geq X)$. Since $X \geq 0$, it follows that

$$\text{not } (0^L \geq X) \text{ or not } (0 \geq X^R).$$

But there is no option 0^L, so, in effect, all this tells us is that

$$\text{not } (0 \geq X^R). \tag{1}$$

Next, since not $(0 \geq X)$, it follows that either

$$X^L \geq 0,$$

for some option X^L or

$$X \geq 0^R,$$

for some option 0^R. But there is no option 0^R, so

$$X^L \geq 0, \tag{2}$$

for some option X^L.

According to (1), no right option of X is either equal to or less than zero. In other words every X^R is either positive or fuzzy. Inductively applying the conclusion of the theorem to X^R, it follows that in every X^R there is either a winning strategy for Left or a winning strategy for the first player. Suppose that we now play the game X as Left but Right has the first move. Right moves first and picks an option X^R. As Left we will make the first move in this option. But we have just established that in X^R there is either a winning strategy for Left or a winning strategy for the first player. As Left, we have a winning strategy in X^R. Thus Left has a winning strategy in the game X if Right goes first.

Pursuing the information given by (2), there is a left option X^L that is either positive or equal to zero. Applying the conclusion of the theorem inductively to X^L, we conclude that X^L has a winning strategy for Left, or a winning strategy for the second player. Suppose that we now play the game X as Left and have first move. If we're smart, we'll pick the option X^L. Right makes the first move in this option and, as Left, we make the second. But in X^L there is either a winning strategy for left or a winning strategy for the second player. As Left, we have a winning strategy in the game X^L. Thus Left has a winning strategy in the game X if Left goes first.

7.2. The Preferential Ordering of Games

Putting together the conclusions of the last two paragraphs, Left has a winning strategy in X no matter who goes first.

This argument can be reversed in order to deduce that if Left has a winning strategy in X, then X is a positive game. $\square$

Corollary. *Every game X either has a winning strategy for the first player, a winning strategy for the second player, a winning strategy for Left, or a winning strategy for Right.*

Problems

9) Classify these Hackenbush positions as positive, negative, zero, or fuzzy.

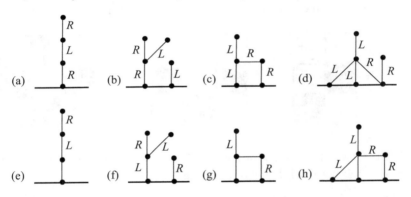

(*Hint*: Play the games and see who wins.)

(*Examples*: In the game ![L R graph], each player has only one move so the second player always wins. This is a zero game.

In the game ![L R graph with extra edge], the first player wins by hacking the unlabeled edge. This leaves the game in the position of the previous example, where the first player, now playing second, wins. This game is fuzzy. Notice how you build knowledge of a Hackenbush game edge by edge.

10) Let $X = \{A, B, C, \ldots | D, E, F, \ldots\}$ and suppose that $A \leq B$. Then A is called a **dominated** option (Left will prefer B). Similarly if $D \leq E$, then E is dominated (Right will prefer D). Prove that dominated options can be eliminated, that is X equals the game $\{B, C, \ldots | D, F, \ldots\}$.

11) Classify these Col positions as positive, negative, zero, or fuzzy.

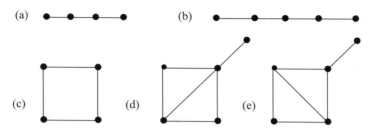

12) Classify the positions in problem 11 as positions in Snort.

13) Classify these Dominos positions as positive, negative, zero, or fuzzy.

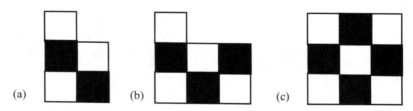

14) Complete the proof of the classification theorem (a) by showing that X is positive if Left has a winning strategy.

15) Prove the classification theorem (c).

Summary

Combinatorial games have an ordering by preference in which $X \geq Y$ means that the game X offers some advantage to Left. Using this order, games can be classified into four categories depending on whether Left, Right, the first player, or the second player has a winning strategy.

7.3 The Arithmetic of Games

Addition of games

Addition of games has to do with playing several games at once. The definition makes this precise.

7.3. The Arithmetic of Games

Definition. If X and Y are games, then the sum $X + Y$ is the game

$$X + Y = \{X^L + Y, X + Y^L \mid X^R + Y, X + Y^R\}.$$

The options of $X + Y$ describe how simultaneous play of X and Y goes. With each turn, the active player moves in X or in Y but not in both. If Left starts, the legal moves are either to a left option of X (leaving the game Y alone) or to a left option of Y (leaving the game X alone). In the first case, Right must either respond to Left's move in X or begin play in Y, that is, right has been left with the position $X^L + Y$. In the second case (if Left moves in Y), the position left for Right is the sum of X (left alone) and some left option of Y, that is $X + Y^L$.

This explains the left options in $X + Y$. A similar discussion justifies the definition of the right options of $X + Y$. The sum ends, as usual, when a player can't move, that is, has no moves in either X or Y.

Proof by playing the game

The classification theorem creates a proof technique unique to Conway's theory. Equations and inequalities can be proved, often rather easily, by playing a game.

Suppose we wish to verify an equation involving games and sums of games. We first move all terms of the equation to one side of the equals sign so that the proof amounts to proving that a certain game equals zero. Then play the game. The game is the zero game if the second player has a winning strategy. If so, the equation is proven.

Problems

1) Verify these identities.

   (a) $0 + 0 \equiv 0$

   (b) $1 + 0 \equiv 1$

   (c) $* + 0 \equiv *$

   (d) $1 + 1 \equiv 2$

2) Prove these by playing the game.
   (a) $2 + (-1) \geq 0$
   (b) $2 + (-1) + (-1) = 0$
   (c) $* + * = 0$
   (d) $1/2 + 1/2 + (-1) = 0$

   (*Example*: Prove that $1 + (-1) = 0$. Since $1 = \{0|\ \}$ and $-1 = \{\ |0\}$, each player can move in only one of these games. The second player, Left or Right, must make the last move and so wins. This proves that $1 + (-1)$ is a zero game.)

3) Prove that $X + 0 \equiv X$ for all games X.

4) If Y is a zero game, use a game playing argument to show that the outcome of the game $X + Y$ is the same as the outcome of the game X.

5) Addition of games appears naturally in Nim, where every position is the sum of positions based on a single heap of coins. Under what circumstances do examples of addition arise in Hackenbush, Dominos, Col, and Snort.

Properties of addition

The commutative and associative laws are more or less obvious. Note that they are identities, not just equalities.

Theorem 7.3.1. *For all games X, Y, and Z, $X + Y \equiv Y + X$, and $(X + Y) + Z \equiv X + (Y + Z)$.*

Theorem 7.3.2. *For all games X, Y, and Z, if $X \geq 0$ and $Y \geq 0$, then $X + Y \geq 0$.*

Problems

6) Prove Theorem 7.3.1.

7) Prove Theorem 7.3.2.

   (*Note*: At least two proofs are possible: one using infinite descent and a direct proof based on playing the game $X + Y$. Find both.)

7.3. The Arithmetic of Games

Negation

A game is **impartial** if Left and Right have the same options in every position. Otherwise the game is **partisan**. Nim, for example, is impartial; Dominos is partizan. Starting with a partizan game, we get a different game if we systematically switch the options of Left and Right throughout. This produces the negative of the game.

Definition. For any game X, the **negative** of X (written $-X$) is the game defined by
$$-X = \{-X^R | -X^L\}.$$

The following theorems outline familiar properties of negation.

Theorem 7.3.3. *For any games X and Y*

(a) $-(X + Y) \equiv -X + -Y$,

(b) $-(-X) \equiv X$, *and*

(c) $X \geq Y$ *if and only if* $-Y \geq -X$.

Theorem 7.3.4. *For any games X and Y, $X \geq Y$ if and only if $X - Y \geq 0$.*

Theorem 7.3.5. *For any game X, $X + (-X) = 0$.*

Thus, $-X$ is an additive inverse for X.

Problems

8) Prove Theorem 7.3.3.

9) Prove Theorem 7.3.4.

    (*Note*: Find two proofs, one by infinite descent and another by playing the game $(X - Y)$.)

10) Prove Theorem 7.3.5.

    (*Note*: Find three proofs: one by infinite descent, one by playing a game, and one by applying a theorem already proven.)

11) Prove that "equals added to equals are equal" is true of the arithmetic of games, that is, if $X = Y$ and $W = Z$, then $X + W = Y + Z$.

12) Explain why impartial games are their own negatives. Conclude that an impartial game is either fuzzy or zero.

13) Define

$$1* = \{0|0\}, \quad *2 = \{0, *1|0, *1\}, \text{ and } *3 = \{0, *1, *2|0, *1, *2\},$$

and more generally

$$\{*n = \{0, *1, \ldots, *(n-1)|0, *1, \ldots, *(n-1)\}.$$

Prove that the sum of two games of the form $*n$ equals another of the same form.

Summary

Combinatorial games have an order relation and an addition operation that satisfy most of the usual axioms of order and addition: order is transitive and reflexive, addition is associative and commutative; there is a zero game (additive identity) and all games have an additive inverse. However, some pairs of games are not comparable so the ordering of games is only a partial order.

7.4 The Surreal Numbers

The surreal numbers constitute a special class of combinatorial games. They form an ordered field that is an extension of the real numbers.

Conway's theory resembles Dedekind's theory of the reals. Recall (from Chapter 2) that Dedekind defined a real number as a cut. A cut divides the rational numbers into two sets, one containing numbers greater than all the numbers in the other, the other containing the rational numbers less than all the numbers in the first. The surreal numbers generalize this idea; every surreal is defined by a cut not among a fixed set of numbers but among the surreal numbers already created, a class that is constantly evolving.

Definition. A **surreal number** is a game x in which

(a) all the options of x are numbers, and

(b) no inequality of the type $x^L \geq x^R$ occurs, that is, no left option of x is greater than or equal to a right option.

The set of all surreal numbers will be notated $\mathbb{S}$.

7.4. The Surreal Numbers

This definition assumes that some surreal numbers are already defined. There is no particular problem about this. Like a game, each number is defined on a certain day. The difference between numbers and games is that numbers only have other numbers as left and right options, and no left option can be greater than or equal to any right option. Proofs about numbers use infinite descent as do proofs about games.

On day zero, the surreal number $0 = \{\,|\,\}$ is born. Parts (a) and (b) of the definition of surreal number are trivially satisfied. Other examples of numbers born in the first few days are

$$1 = \{0|\,\} \qquad 2 = \{0,1|\,\} \qquad 3 = \{0,1,2|\,\}$$
$$-1 = \{\,|0\} \qquad -2 = \{\,|0,-1\} \qquad -3 = \{\,|0,-1,-2\}$$
$$1/2 = \{0|1\} \qquad 1/4 = \{0|1/2\} \qquad 3/4 = \{1/2|1\}$$

and

$$-1/2 = \{-1|0\} \qquad -1/4 = \{-1/2|0\} \qquad -3/4 = \{-1|-1/2\}.$$

In general, the integers, positive and negative, are defined by

$$n = \{0, 1, \ldots, (n-1)|\,\},$$

and

$$-n = \{\,|0, -1, -2, \ldots, -(n-1)\},$$

and are defined on day n. It is fairly easy to show that their addition is the same as the usual addition of integers (see problem 1). Meanwhile, let's verify that the operations we have defined for games apply to numbers.

Theorem 7.4.1. (a) *If* $\mathbf{x}$ *is a surreal number, then so is* $-\mathbf{x}$,

(b) *If* $\mathbf{x}$ *and* $\mathbf{y}$ *are numbers, then so is* $\mathbf{x} + \mathbf{y}$,

(c) *If* $\mathbf{x}$ *and* $\mathbf{y}$ *are numbers, then either* $\mathbf{x} \geq \mathbf{y}$ *or* $\mathbf{x} \leq \mathbf{y}$.

Thus the surreal numbers are closed under addition and formation of additive inverses. Furthermore, it follows from (c) that they satisfy the trichotomy law and hence are linearly ordered (see problem 3).

Problems

1) Verify that $2 + 2 = 4, 2 + 3 = 5$, etc.

2) Prove parts *a* and *b* of Theorem 7.4.1.

3) Prove that the surreal numbers satisfy the trichotomy law.

4) Prove that if **x** is a number, then
$$\mathbf{x}^L < \mathbf{x} < \mathbf{x}^R.$$

5) Find all surreal numbers born on days 0, 1, 2 and 3. The birth of these numbers follows a pattern. Find it.

Surreal multiplication

Multiplication of surreals is considerably more complicated than addition, because it does not have a natural extension to games. Surreal multiplication is for surreal numbers only.

To introduce multiplication, begin by reconsidering addition. Let **x** and **y** be numbers. In order that $\mathbf{x} + \mathbf{y}$ be a number, the left options of $\mathbf{x} + \mathbf{y}$ must be numbers that are smaller than $\mathbf{x} + \mathbf{y}$, and the right options must be numbers that are larger than $\mathbf{x} + \mathbf{y}$. This suggests that the numbers

$$\mathbf{x}^L + \mathbf{y} \text{ and } \mathbf{x} + \mathbf{y}^L$$

be taken as left options for $\mathbf{x} + \mathbf{y}$. They are smaller than $\mathbf{x} + \mathbf{y}$ since $\mathbf{x}^L$ is smaller than **x** and $\mathbf{y}^L$ is smaller than **y**. In a similar fashion the right options of $\mathbf{x} + \mathbf{y}$ could be

$$\mathbf{x}^R + \mathbf{y} \text{ and } \mathbf{x} + \mathbf{y}^R.$$

This motivates the definition

$$\mathbf{x} + \mathbf{y} = \{\mathbf{x}^L + \mathbf{y}, \mathbf{x} + \mathbf{y}^L | \mathbf{x}^R + \mathbf{y}, \mathbf{x} + \mathbf{y}^R\},$$

which works, inductively, because each option involves at least one number that is older than $\mathbf{x} + \mathbf{y}$, i.e., a number born before $\mathbf{x} + \mathbf{y}$.

One is tempted to try the same idea for multiplication, defining

$$\mathbf{xy} = \{\mathbf{x}^L\mathbf{y}, \mathbf{xy}^L | \mathbf{x}^R\mathbf{y}, \mathbf{xy}^R\}.$$

This doesn't work (see problem 5). A more fruitful idea is to observe that since

$$(\mathbf{x} - \mathbf{x}^L) > 0 \text{ and } (\mathbf{y} - \mathbf{y}^L) > 0,$$

we can deduce

$$(\mathbf{x} - \mathbf{x}^L)(\mathbf{y} - \mathbf{y}^L) > 0,$$

7.4. The Surreal Numbers

so that
$$xy > x^L y + xy^L - x^L y^L.$$

This last quantity is a suitable left option for **xy** because the products in it all involve at least one older number.

Definition. If **x** and **y** are numbers, let

$$xy = \{x^L y + xy^L - x^L y^L, x^R y + xy^R - x^R y^R \\ \mid x^L y + xy^R - x^L y^R, x^R y + xy^L - x^R y^L\}.$$

Problems

6) What happens if multiplication is defined

$$xy = \{x^L y, xy^L \mid x^R y, xy^R\}.$$

(*Hint*: What is multiplication by zero?)

7) Calculate these products using the definition:
   (a) $0 \cdot 1$
   (b) $1 \cdot 1$
   (c) $1 \cdot 2$
   (d) $2 \cdot 2$
   (e) $2 \cdot 1/2$

Properties of multiplication

We don't yet know that **xy** is a number. It is a game, however, so we can use results already established for games to prove properties of multiplication:

Theorem 7.4.2. *For all surreal numbers* **x**, **y** *and* **z**

(a) $x0 \equiv 0$

(b) $x1 \equiv x$

(c) $xy \equiv yx$

(d) $(-x)y \equiv x(-y) \equiv -(xy)$

(e) $(x + y)z = xz + yz$

(f) $(xy)z = x(yz)$

Now we can prove that the surreal numbers are closed under multiplication.

Theorem 7.4.3. *If x and y are surreal numbers, then xy is a surreal number. If x and y are positive numbers then xy is positive.*

Problems

8) Prove Theorem 7.4.2.

9) Prove that if $x_1 = x_2$, then $x_1 y = x_2 y$.

10) Prove that if $x_1 = x_2$ or $y_1 = y_2$, then $x_1 y_2 + x_2 y_1 = x_1 y_1 + x_2 y_2$

11) Prove that if $x_1 > x_2$ and $y_1 > y_2$, then $x_1 y_2 + x_2 y_1 > x_1 y_1 + x_2 y_2$

12) Prove that if $x_1 \geq x_2$ and $y_1 \geq y_2$, then $x_1 y_2 + x_2 y_1 \geq x_1 y_1 + x_2 y_2$

13) Prove Theorem 7.4.3.

Summary

The surreal numbers are a peculiar kind of game in which Left doesn't prefer it's own options over Right's and vice versa. Unlike games in general, the surreal numbers possess a second arithmetic operation, multiplication, with the usual algebraic properties of multiplication including a multiplicative identity.

7.5 The Nature of the Surreal Line
One day at a time

Which numbers are surreal? We begin with a theorem that enables us to identify new numbers as they are born.
  Let
$$x = \{a, b, c, \ldots | d, e, f, \ldots\},$$
where $a, b, c, d, e, f, \ldots$, are all numbers and each left option is strictly less than each right option. Then x is a number, but which? Following Conway, let us say the number y **fits** if y is strictly greater than $a, b, c, \ldots$ and strictly less than $d, e, f, \ldots$, that is, if
$$a, b, c, \cdots < y < d, e, f, \ldots.$$

7.5. The Nature of the Surreal Line

The numbers that fit are candidates for **x**. For example, on day 4, the number $\mathbf{x} = \{-1|2\}$ appears. This has to be a number between -1 and 2. According to the next theorem, **x** equals the oldest (i.e., earliest created) number that fits. For $\{-1|2\}$, 0 (the oldest number there is) fits, so $\mathbf{x} = \{-1|2\} = 0$.

Theorem 7.5.1. *(The simplicity theorem). Let* $\mathbf{x} = \{\mathbf{x}^L|\mathbf{x}^R\}$ *be a number. Suppose that the number* **z** *satisfies*

$$\mathbf{x}^L < \mathbf{z} < \mathbf{x}^R,$$

but no option of **z** *(i.e., no ancestor of* **z***) satisfies this condition. Then* $\mathbf{x} = \mathbf{z}$.

The alternative forms of the numbers listed below, are all justified by this theorem.

Day 0: $0 = \{\,|\,\}$

Day 1: $1 = \{0|\,\}$

             $-1 = \{\,|0\}$

Day 2: $0 = \{-1|1\}$ — a new form of 0

             $1/2 = \{0|1\} = \{-1, 0|1\}$

             $-1/2 = \{-1|0\} = \{-1|0, 1\}$

             $2 = \{1|\,\} = \{0, 1|\,\} = \{-1, 0, 1|\,\}$

                $= \{-1, 1|\,\}$

             $-2 = \{\,|-1\} = \{\,|-1, 0\}$

                $= \{\,|-1, 1\} = \{\,|-1, 0, 1\}$

For example, $\{-1, 0|1\}$ has to be a number between 0 and 1 (strictly greater than 0 and strictly smaller than 1). The oldest such number is $1/2$, therefore $\{-1, 0|1\} = 1/2$.

The simplicity theorem also makes it easy to prove addition facts. For example, by definition

$$1/2 + 1/2 = \{1/2 + 0, 0 + 1/2|1/2 + 1, 1 + 1/2\} = \{1/2|1 + 1/2\}.$$

By the simplicity theorem this is the oldest number strictly greater than $1/2$ and strictly less than $1 + 1/2$. This is 1 because the only number older than 1 has to be 0, and 0 doesn't fit. Thus $1/2 + 1/2 = 1$.

On day 3 a lot of numbers are born but they are alternate forms of only a few numbers. The simplicity theorem implies that each day only one number can be born between a neighboring pair of old numbers. Thus we get

Day 3 : $3 = \{2| \ \} = \{1, 2| \ \} = \ldots$
$3/2 = \{1|2\} = \{-1, 1|2\} = \ldots$
$3/4 = \{1/2|1\} = \ldots$
$1/4 = \{0|1/2\} = \ldots$
$-1/4 = \{-1/2|0\} = \ldots$
$-3/4 = \{-1|-1/2\} = \ldots$
$-3/2 = \{-2|-1\} = \ldots$
$-3 = \{ \ |-2\} = \ldots$

Problems

1) Prove the simplicity theorem.

   (*Hint*: Prove separately that $\mathbf{x} \geq \mathbf{z}$ and $\mathbf{x} \leq \mathbf{z}$.)

2) Verify that the preceding list includes all new numbers born on day 3.

3) Verify that the numbers created on day 3 behave as their names indicate they should. That is, show that

$$3 = 2 + 1,$$
$$2(3/2) = 3,$$
$$(3/4) + (3/4) = 3/2,$$
$$1/4 + 1/4 = 1/2,$$

   and so forth.

4) On a finite numbered day, only one number can be born between two adjacent numbers. Prove this.

5) On a finite numbered day, prove that the number of numbers created is twice the number created the previous day. Show that all numbers created on a finite day are **dyadic rationals**, that is, rational numbers whose denominator is a power of 2.

7.5. The Nature of the Surreal Line

Day ω

After all the finite numbered days have passed, there is a next day! Call it ω. On day ω, many new numbers are born. One example is:

$$\mathbf{a} = \{\text{all older numbers, } y, \text{ such that } 3y < 1 \\ |\text{all older numbers, } y, \text{ such that } 1 < 3y\}.$$

Typical left options of $\mathbf{a}$ are $1/4, 5/16$, and $21/64$; typical right options are $1/2, 3/8$. The simplicity theorem implies that $\mathbf{a} + \mathbf{a} + \mathbf{a} = 3$; therefore, $\mathbf{a} = 1/3$. In a similar way, the other real numbers not yet born—rational and irrational—including numbers like π and e—all arrive on day ω.

The largest number born on day ω is the number ✪:

$$✪ = \{0, 1, 2, 3, \ldots | \ \},$$

which is larger than all previously created numbers. Also created is the number

$$-✪ = \{ \ |0, -1, -2, -3, \ldots\}.$$

The smallest positive number created on day ω is

$$\{0|1, 1/2, 1/4, 1/8, \ldots\},$$

which turns out to be $1/✪$.

These numbers are analogous to the infinitely large and infinitely small hyperreals and note that the set of birthdays is the well-ordered set $W = \{0, 1, 2, 3, \ldots, \omega\}$ discussed in Chapter 6.

The creation of numbers does not stop here. On day $\omega + 1$, more numbers are created such as

$$\{0, 1, 2, 3, \ldots, ✪| \ \},$$

and

$$\{0, 1, 2, 3, \ldots |✪\}.$$

And after day $\omega + 1$, there is day $\omega + 2$ and so on, and on.

Problems

6) Prove that $\mathbf{a} + \mathbf{a} + \mathbf{a} = 3$.

7) Show that $\sqrt{2}$ is created on day ω.

8) Show that
$$\{0, 1, 2, 3, \ldots, \omega |\ \} = \omega + 1,$$
and
$$\{0, 1, 2, 3, \ldots | \omega\} = \omega - 1.$$

9) Show that
$$\{0, 1, 2, 3, \ldots | \omega, \omega - 1\} = \omega - 2,$$
and in general
$$\{0, 1, 2, \ldots | \omega, \omega - 1, \omega - 2, \ldots, \omega - (n-1)\} = \omega - n.$$

10) On what day is
$$z = \{0, 1, 2, \ldots | \omega, \omega - 1, \omega - 2, \ldots\}$$
created? Show that $z = \omega/2$.

Summary

The surreal numbers include the real numbers and certain infinite and infinitesimal numbers. Unlike the hyperreals, these numbers are not created all at once but step-by-step, one day at a time. In order to reach infinitely large and infinitesimal surreals, it is necessary to carry out this process to infinity and beyond.

7.6 More Surreal Numbers

Completing the construction of the surreals

We have almost finished building the surreals. The missing ingredient is the existence of multiplicative inverses. To begin we need a lemma.

Lemma. *Each positive surreal x has a form in which 0 is a left option and every other left option is positive.*

Definition. Let x be a positive surreal. Let x be given in the form described in the previous lemma. Then, the **reciprocal** of x is defined by setting

$$y = \left\{ 0, \frac{1 + (x^R - x)y^L}{x^R}, \frac{1 + (x^L - x)y^R}{x^L} \ \middle|\ \frac{1 + (x^L - x)y^L}{x^L}, \frac{1 + (x^R - x)y^R}{x^R} \right\}.$$

7.6. More Surreal Numbers

Like other definitions in the theory of games, this is recursive: it supposes that reciprocals have been already defined for the positive right and left options of **x**. To start the induction, **x** must have at least one positive right or left option. This means that **x** must not be zero, but, of course, zero has no reciprocal.

Moreover, this definition is doubly recursive: it defines the options of **y** not only in terms of options of **x**, but also in terms of previously created options of **y**. The second induction is started by assuming that 0 is a left option of **y**.

For example let $\mathbf{x} = 3 = \{0, 1, 2|\}$, and consider the computation of $\mathbf{y} = 1/3$. According to the definition,

$$\mathbf{y} = \{0, \ldots | \ldots \},$$

where further right and left options of **y** remain to be calculated by the four fractional formulas above. Each uses an option of **y**. So far there is only one known option of **y**: the left option, $\mathbf{y}^L = 0$. As it happens, **x** only has left options and, of the two fractional formulas that use $\mathbf{y}^L$, only one uses $\mathbf{x}^L$. Thus we must start by using the formula

$$\frac{1 + (\mathbf{x}^L - \mathbf{x})\mathbf{y}^L}{\mathbf{x}^L}.$$

With $\mathbf{y}^L = 0$ and $\mathbf{x}^L = 2$, and assuming (inductively) that the inverse of $\mathbf{x}^L$ has already been constructed and is the familiar number $1/2$, we get

$$\mathbf{y} = \left\{ 0, \ldots \left| \frac{1 + (2-3)0}{2}, \ldots \right. \right\} = \left\{ 0, \ldots \left| \frac{1}{2}, \ldots \right. \right\}.$$

We now have a right option $\mathbf{y}^R = 1/2$. Using this option and $\mathbf{x}^L = 2$ again (we could use 1, but 2 is more interesting), we get

$$\mathbf{y} = \left\{ 0, \frac{1 + (2-3)\frac{1}{2}}{2}, \ldots \right\} = \left\{ 0, \frac{1}{4}, \ldots \left| \frac{1}{2}, \ldots \right. \right\}.$$

Continuing in this way, we obtain further options for **y**:

$$\mathbf{y} = \left\{ 0, \frac{1}{4}, \frac{5}{16}, \frac{21}{64}, \ldots \left| \frac{1}{2}, \frac{3}{8}, \frac{11}{32}, \ldots \right. \right\}.$$

The process never ends because **y** has an infinite number of left and right options. Nevertheless, **y** satisfies the basic rule of games: there is no infinite sequence of positions of **y** each one an option of its predecessor. In fact, one move in **y** by either right or left leads to a position with only a finite number of options. Observe that the options we have calculated so far do appear to approximate $1/3$ better and better, both from above and from below.

Problems

1) Prove the Lemma.

   (*Hint*: Use the result on dominated options in problem 10 of section 7.2.)

2) Extend the computation of $1/3$ to obtain one more left and one more right option.

3) Use the definition to calculate a few left and right options of the reciprocals of these numbers:

   (a) $2 = \{0, 1|\ \}$
   (b) $5 = \{0, 1, 2, 3, 4|\ \}$
   (c) $1/2 = \{0|1\}$.

   (*Hint*: In each case, assume that reciprocals are already calculated for the given options and that they behave as expected. Thus to calculate $1/5$, assume the reciprocal of 3 is $1/3$ and the reciprocal of 4 is $1/4$.

4) What is the base case of the definition of multiplication?

The surreals are a field

The next theorem, whose proof is intricate, states that the reciprocals just defined are multiplicative inverses.

Theorem 7.6.1. *Let* **x** *be a positive surreal. Suppose that* **y** *is the reciprocal of* **x**. *Then*

(a) $\mathbf{xy}^L < 1 < \mathbf{xy}^R$

(b) **y** *is a number*

(c) $(\mathbf{xy})^L < 1 < (\mathbf{xy})^R$

(d) $\mathbf{xy} = 1$

   This allows us to complete the construction of the surreals.

7.6. More Surreal Numbers

Theorem 7.6.2. *The surreal numbers are a linearly ordered field containing the real numbers as well as infinite and infinitesimal numbers.*

The proofs of Theorems 7.6.1 and 7.6.2 are complicated. See Conway [J4] for the details.

Roots

Square roots are defined by a formula similar to that used for reciprocals: let **x** be a positive surreal, then define

$$\sqrt{\mathbf{x}} = \left\{ \sqrt{\mathbf{x}^L}, \frac{\mathbf{x} + \mathbf{y}^L \mathbf{y}^R}{\mathbf{y}^L + \mathbf{y}^R} \mid \sqrt{\mathbf{x}^R}, \frac{\mathbf{x} + \mathbf{y}^L \mathbf{y}^{L'}}{\mathbf{y}^L + \mathbf{y}^{L'}}, \frac{\mathbf{x} + \mathbf{y}^R \mathbf{y}^{R'}}{\mathbf{y}^R + \mathbf{y}^{R'}} \right\},$$

where some options of **y** require using two options that have been already obtained (and chosen so that none of the denominators are zero). The process is started by including the square roots of the options of **x**.

Problems

5) Compute some options for the square roots of $2 = \{0, 1|\}$ and $3 = \{0, 1, 2|\}$.

6) Prove that $\sqrt{\mathbf{x}}$ as defined is a surreal number

7) Prove that $\sqrt{\mathbf{x}}$ as defined is a square root of **x**.

Summary

In this section the final steps in the construction of the surreals were described. The situation is analogous to the problems encountered in Dedekind's construction of the reals (Chapter 2). Dedekind's definition made each real, like a surreal, a partition of numbers into two sets, a left set and a right set. Dedekind's partitions are called cuts, and the left and right sets are subsets of the rationals. In Dedekind's construction of the reals and in Conway's construction of the surreals, the fundamental properties of multiplication are difficult to verify.

7.7 Analyzing Games with Numbers

As games, numbers are boring. Because $x^L < x < x^R$, a move by either player in x decreases the value of the game for that player. Playing a surreal number, neither player wants to move since each prefers the other's options. In a sum of games, $G + x$, if x is a number and G is not, each player tends to move in G rather than in x. Numbers nevertheless appear automatically in some games as the next problems show.

Problems

1) What numbers are represented by these Hackenbush positions:

2) Let P represent a Hackenbush position. Then the value of the position $f(P)$ as shown in the figure below depends only on the value of P.

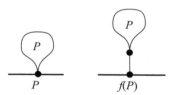

Fill in the missing entries in this table:

value of P	-4	-3	-2	-1	$-\frac{1}{2}$	0	$\frac{1}{2}$	1	2	3
value of $f(P)$				$1/2$		1				

3) A game of Hackenbush in which all edges are labeled L or R is called **partizan Hackenbush**. In partizan Hackenbush every edge can be hacked by Left, or Right, but not by both. Prove that every partizan Hackenbush position is a number.

(*Hint*: Establish (simultaneously) the fact that a move by Left decreases the value of the game and every move by Right increases the value of the game.)

4) If x is a position in Col, then
$$X^L + * \geq X \geq X^R + *$$
Establish this by playing the games $X - X^L - *$ and $X^R + * - X$.

7.7. Analyzing Games with Numbers

5) Prove that every position in Col is either a number or of the form

$$x + * = \{x|x\},$$

where **x** is a number.

6) Evaluate these positions in Col.

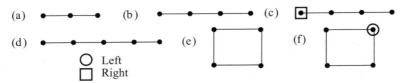

7) Find examples of positions that are numbers in Dominos and Snort.

Stopping values

In the play of a game, if a player moves to a position that is a number, play may as well stop because neither player wants to continue. This number can be used to score the game.

Definition. Let G be a game. Any number that can arise as a position during play of G is called a **stopping value** of G. The remaining positions of G are **active positions**.

If numbers are used to score a game, Left will play so as to maximize the position at which play stops; Right will play to minimize the score. These maximum and minimum scores can be defined inductively as follows:

Definition. A game is called **short** if it has only a finite total number of options. Let G be a short game. If G is not a number, set

$$L(G) = \max_{G^L} R(G^L),$$

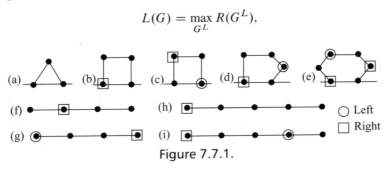

Figure 7.7.1.

and
$$R(G) = \min_{G^R} L(G^R).$$

If G is a number, set
$$L(G) = R(G) = G.$$

The quantities, $L(G)$ and $R(G)$, are called, respectively, the **left stopping value** and the **right stopping value** of G.

The restriction to short games is needed so that the maximum and minimum exist. The left and right stopping values of G can now be used to determine how G compares with all numbers.

Theorem 7.7.1. *A short game G is greater than all numbers $\mathbf{x} < R(G)$, less than all numbers $\mathbf{x} > L(G)$, and confused with numbers $\mathbf{x}$ such that $R(G) < \mathbf{x} < L(G)$.*

Problems

8) Prove Theorem 7.7.1.

   (*Hint*: Use proof by induction to play $G + \mathbf{x}$ under the various conditions described in the theorem.)

9) Find $R(G)$ and $L(G)$ for the Col positions in Figure 7.7.1.

10) Find $R(G)$ and $L(G)$ for the Snort positions in Figure 7.7.2.

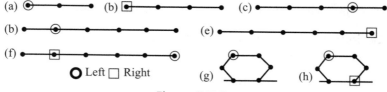

Figure 7.7.2.

11) Find $R(G)$ and $L(G)$ for the Dominos positions in Figure 7.7.3.

12) Find $R(G)$ and $L(G)$ for the Hackenbush positions in Figure 7.7.4.

13) The game of **Hex** is played on an $n \times n$ square board of dots (Figure 7.7.5). Players take turns marking one dot as in Col and Snort. Left's goal is to create a path of dots from the top of the board to the bottom;

7.7. Analyzing Games with Numbers

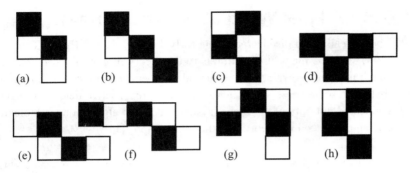

Figure 7.7.3.

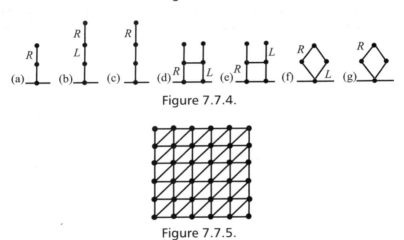

Figure 7.7.4.

Figure 7.7.5.

Right's goal is to connect the vertical sides. Investigate this game using the theory developed in this Chapter.

(*Hint*: Begin with small boards.)

Summary

Numbers appear in all combinatorial games as final positions, positions where play of the game will likely stop because neither player wants to continue. The surreals, therefore, help evaluate the endings of games. In particular the left and right stopping values of a game delineate the range of possible numerical outcomes of a game.

7.8 A Final Word about the Surreals

The surreals were discovered in the early 1970's by John Horton Conway (1937–). Their major application is to the theory of combinatorial games. Much more is known about the surreal numbers and combinatorial games than is presented here. Combinatorial games are an active field of research, connected not only with number systems, but also algorithmic complexity and algebraic coding theory (see the bibliography [J5]).

General References for the surreals. Conway's theory of games first appeared in [J4], which this treatment largely follows. The multi-volume set [J1] is more recent and more comprehensive. The surreals apart from their connection with games are described in the novel [J7] and also in [J6].

Bibliography

General References

[A1] Conway, John H., and Richard Guy, *The Book of Numbers*, Springer, 1995.

[A2] Ebbinghaus, H.-D., et. al., *Numbers* (H. L. S. Orde, trans.), Springer, 1990.

[A3] Kline, Morris, *Mathematical Thought from Ancient to Modern Times* (3 vols.), Oxford University Press, 1990.

How to Find Proofs

[B1] Franklin, James and Albert Daoud, *Introduction to Proofs in Mathematics*, Kew Books, 2011.

[B2] Lakatos, Imre, *Proofs and Refutations: The Logic of Mathematical Discovery*, Cambridge, 1976.

[B3] Pólya, Georg, *How To Solve It: A New Aspect of Mathematical Method*, Ishi Press, 2009.

[B4] Solow, Daniel, *How to Read and Do Proofs: An Introduction to Mathematical Thought Processes* (5th ed.), Wiley, 2010.

[B5] Velleman, Daniel, *How to Prove It: A Structured Approach* (2nd ed.), Cambridge, 2006.

Algebra and Number Systems in General

[C1] Halmos, Paul R., *Naive Set Theory*, Springer, 1998.

[C2] Herstein, I. N., *Topics in Algebra* (2nd ed.), Wiley, 1975.

[C3] Kantor, I. L. and A. S. Solodovnikov, *Hypercomplex Numbers: An Elementary Introduction to Algebras* (A. Shenitzer, trans.), 1989.

[C4] Littlewood, D. E., *A University Algebra*, Dover, 1971.

[C5] MacLane, Saunders and Garrett Birkoff, *Algebra*, AMS/Chelsea, 1999.

[C6] van der Waerden, B. L., *Modern Algebra*, Springer, 2003.

[C7] Weber, Heinrich Martin, *Lehrbuch der Algebra*, v. 1, (3rd ed., in German), Nabu Press, 2010.

Number Theory

[D1] Cox, David A., *Primes of the form $x^2 + ny^2$: Fermat, Class Field Theory and Complex Multiplication*, Wiley & Sons, 1997.

[D2] Hardy, Geoffrey, E. M. Wright, and Andrew Wiles, *An Introduction to the Theory of Numbers* (6th ed., Roger Heath-Brown and Joseph Silverman, eds.), Oxford, 2008.

[D3] Rosen, Kenneth, *Elementary Number Theory and its Applications* (6th ed.), Addison-Wesley, 2010.

The Real Numbers

[E1] Apostol, Tom M., *Mathematical Analysis* (2nd ed.), Addison-Wesley, 1973.

[E2] Burrill, Claude W., *Foundations of Real Numbers*, McGraw-Hill, Inc., 1967.

[E3] Cohen, Leon W., and Gertrude Ehrlich, *The Structure of the Real Number System*, Van Nostrand, 1963.

The Complex Numbers

[F1] Ahlfors, Lars V., *Complex Analysis* (3rd ed.), McGraw-Hill, 1979.

[F2] Brown, James Ward and Ruel V. Churchill, *Complex Variables and Applications* (8th ed.), McGraw-Hill, 2008.

[F3] Greenleaf, Frederick P., *Introduction to Complex Variables*, W. B. Saunders, Philadelphia, 1972.

[F4] Needham, Tristram, *Visual Complex Analysis*, Oxford, 1999.

The Quaternions

[G1] Abonyi, I., "Quaternion Representation of Lorentz Group for Classical Physics," *J. Phys. A. Math. & Gen.* **24** (July 1991) 3245–3254.

[G2] Althoen, S.C., et al., "Rotational Scaled Quaternion Division Algebras," *J. Algebra* **146** (1992) 124–143.

[G3] Coxeter, H.S.M., "Quaternions and Reflections," *Am. Math. Monthly* **53** (1946) 136–146.

[G4] Deavours, C.A., "The Quaternion Calculus," *Am. Math. Monthly* **80** (1973) 995–1008.

[G5] Doran, Chris and Althony Lasenby, *Geometric Algebra for Physicists*, Cambridge, 2003.

[G6] Fenchel, Werner, *Elementary Geometry in Hyperbolic Space*, de Gruyter, 1989.

[G7] Gsponer, Andre and Jean-Pierre Hurni, *Quaternions in Mathematical Physics*; available at `arxiv.org/abs/mathph/0510059`, last revised 2008.

[G8] Kuipers, Jack B., *Quaternions and rotation Sequences: a Primer with Applications to Orbits, Aerospace, and Virtual Reality*, Princeton, 1999.

The Constructive Real Numbers

[H1] Bishop, Errett, and Douglas Bridges, *Constructive Analysis*, Springer, 1985.

[H2] Bridger, Mark, *Real Analysis: A Constructive Approach*, Wiley-Interscience, 2007.

[H3] Bridges, Douglas and Fred Richman, *Varieties of Constructive Mathematics*, Cambridge, 1987.

[H4] George, Alexander and Daniel J. Velleman, *Philosophies of Mathematics*, Wiley-Blackwell, 2001.

[H5] Richman, Fred, "Existence Proofs", Am. Math. Monthly **106** (1999) 303–308.

[H6] Schechter, Eric, "Constructivism is Difficult," Am. Math. Monthly **108** (2001) 50–54.

[H7] Troelstra, A.S. and D. van Dalen, *Constructivism in Mathematics: An Introduction*. North-Holland, 1988.

[H8] Velleman, Dan, "Constructivism Liberalized," *Philosophical Review* **102** (1993) 59–84.

The Hyperreal Numbers

[I1] Cutland, Nigel, Mauro di Nasso and David A. Ross (editors), *Nonstandard Methods and Applications in Mathematics*, A.K. Peters/CRC, 2006.

[I2] Henle, James, and Eugene Kleinberg, *Infinitesimal Calculus*, Dover, 2003.

[I3] Hoskins, R.F., *Standard and Nonstandard Analysis: Fundamental Theory, Techniques and Applications*, Horwood, 1990.

[I4] Komjath, Peter and Vilmos Totik, "Ultrafilters," Am. Math. Monthly **115** (2008) 33–44.

[I5] Robert, Alain, *Nonstandard Analysis*, Dover, 2011.

The Surreal Numbers

[J1] Berlekamp, Elwyn, John H. Conway, and Richard K. Guy, *Winning Ways For Your Mathematical Plays* (2nd ed., 4 vols.), A.K. Peters/CRC, 2001–2004.

[J2] Berlekamp, Elwyn, *The Dots and Boxes Game: Sophisticated Child's Play*, A K Peters/CRC, 2000.

[J3] Berlekamp, Elwyn and David Wolfe, *Mathematical Go: Chilling Gets the Last Point*, A.K. Peters/CRC, 1997.

[J4] Conway, J. H., *On Numbers and Games* (2nd ed.), A.K. Peters/CRC, 2000.

[J5] Fraenkel, Aviezri, "Combinatorial Games: Selected Bibliography with a Succinct Gourmet Introduction," *Electronic Journal of Combinatorics*; available at www.combinatorics.org/.

[J6] Gonshor, Harry, *An Introduction to the Theory of Surreal Numbers*, Cambridge, 1986.

[J7] Knuth, Donald E., *Surreal Numbers*, Addison-Wesley, 1974.

Applications

[K1] Deetz, Charles and Oscar Adams, *Elements of Map Projection*, U.S. Department of Commerce, Government Printing Office, 1945.

[K2] Feynman, Richard P., Robert B. Leighton and Matthew Sands, *The Feynman Lectures on Physics*, Addison Wesley, 1970.

[K3] Lorimer, Peter, *The Special Theory of Relativity for Mathematics Students*, World Scientific, 1990.

[K4] Misner, Charles W., Kip S. Thorne, and John Archibald Wheeler, *GRAVITATION*, Freeman, 1973.

[K5] Taylor, Edwin, and John Archibald Wheeler, *Spacetime Physics* (2nd. ed.), Freeman, 1992.

Index

absolute value
   and distance, 20
abstraction, 99
addition
   of fractions, 9
   of games, 184–186
   of inequalitites, 20
   of reals, 12
additive
   identity, 13
   inverse, 7, 13
algebra
   fundamental theorem of, 66
   of the complex numbers, 57
anti-symmetry, 162
applications
   airplane wings, 76
   animation, 94
   automobile bodies, 76
   submarine hulls, 76
   theory of relativity, 86, 89
Archimedean field, 28
Archimedean property, 27, 31, 32, 34, 152
   for the complex numbers, 61
   for the constructive reals, 119
argument, 60
associative law, 12
   for games, 186
   for surreal numbers, 191
axioms
   categorical, 1, 35
   completeness, 26
   for a field, 12, 135
   for equivalence, 3
   for linear order, 18
   for partial order, 162
   for well-ordered set, 163, 164
axis of rotation, 83

Banach, Stefan, 168
Banach-Tarski paradox, 168, 169
big sets, *see* ultrafilter
Bishop, Errett, 97, 103, 124
Bolzano, Bernard, 11
bounded, 29
   above, 25
   below, 26
boundedness theorem
   classical, 50, 51

constructive view of, 121, 122
hyperreal proof, 160
Brouwer, Luitzen, 101, 124

calculus, *see* differential calculus
cancellation law
  of addition, 15
  of multiplication, 15
Cantor reals, 36–42, 48, 49, 98
  addition of, 40
  completeness of, 42
  definition of, 37
  multiplication of, 40
  positive, 41
Cantor, Georg, 36, 43, 53
Cartesian form
  for complex numbers, 58
  for quaternions, 78
categorical axiom system, 1, 12, 35
Cauchy completeness, 28–32
  of the constructive reals, 118
Cauchy sequence, 28–32, 37, 38, 40
  complex numbers, 61
  constructive version of, 98, 109
  hyperreal definition of, 157
  of complex numbers, 61
  positive, 41
Cauchy, Augustin-Louis, 11, 29, 53
Cauchy-Riemann equations, 73
chain rule, 52, 161
  complex version, 72
  real proof, 52
classical logic, 103–106
  conjunction, 104
  disjunction, 104
  existence statement, 104
  implication, 104
  negation, 104

universal statement, 104
classical mathematics, 101
classification theorem, 181, 182
Clifford algebra, 94
cloud, 152
cofinite sets, 139
Col, 176, 177, 184, 186, 201, 202
combinatorial games, 171–188
  active positions, 201
  addition of, 184–186
  classification theorem of, 181, 182
  confused, 179
  dominated options in, 183
  end game, 173
  equality of, 180
  examples, 175, 176
  fuzzy, 181
  incomparable, 179
  initial position of, 172
  left and right options of, 172
  negative of, 187
  ordering of, 178
  position in, 172
  short, 201
  stopping value of, 201
  zero game, 173
commutative law, 13, 94
  for games, 186
completeness, 12, 24–35, 42, 53, 55, 100, 135
  Cauchy, 28, 29
  for the complex numbers, 61, 62
  for the constructive reals, 118, 119
  order, 24–29, 31–33
complex exponential, 62, 63
complex functions, 70–72

Index

211

applications of, 75, 76
Cauchy-Riemann equations, 73
conformality of, 74
derivative, 72–75
Laplace's equation, 73
quadrupole field, 76
real and imaginary parts of, 73
square root, 75
complex numbers, 11, 57–66, 76
addition of, 57
algebra of, 57
Archimedean property, 61
argument of, 60
Cartesian form, 58
completeness of, 61, 62
conjugate of, 60
differential calculus of, 72–75
distance function, 61
imaginary part of, 58
limits of, 61, 72
modulus of, 60
multiplication of, 57
order of, 59
polar form of, 62
quadrupole field, 76
real part of, 58
square roots of, 75
uniqueness, 66
complex quaternions, 86–88, 107
conjugate, 86
Lorentz transformations, 87
norm, 86
polar form, 87
scalar part, 86
vector part, 86
computation, 97
conformality, 74
confused games, 179
conjugate, 60

of a complex quaternion, 86
of a quaternion, 78
conjunction
in classical logic, 104
in constructive logic, 106
construction
of the Cantor reals, 36
of the constructive reals, 109
of the Dedekind reals, 43
of the hyperreals, 138
of the reals, 36
of the surreals, 188
constructive functions, 108
contrast with operations, 108
constructive logic, 106, 108
conjunction, 106
disjunction, 106
existence statement, 107
implication, 106
negation, 106
universal statement, 107
constructive mathematics, *see* constructivism
constructive real numbers, 11, 43, 97, 109–119, 124
addition, 112
alternative definition of equality, 110
Archimedean property of, 119
as regular sequences, 109
boundedness theorem, 121, 122
Cauchy completeness of, 118
completeness of, 118, 119
computation, 97
continuity for, 119, 120
criticism of classical mathematics, 97, 99
differential calculus of, 122
equality, 109

intermediate value theorem, 120
linear order, 114–116
maximum value theorem, 122
multiplication, 112
multiplicative inverses, 117
non-negative, 114
positive, 114
square roots, 122
trichotomy, 116
constructive sets, 108
constructivism, 97–101, 103
continuity
hyperreal version, 159
continuous function
$\varepsilon - \delta$ definition, 51
constructive version, 120
sequence definition, 50
contrapositive proof, 105
Conway, John, 171, 173, 204
cross product, 80, 81
CrossScram, *see* Dominos
curl, 92
cut
addition of, 44
Dedekind, 43, 188
non-negative, 45
null, 44, 45
positive, 45
rational, 44

De Morgan's laws, 105
Dedekind cut, 43
Dedekind reals, 43–49
addition of, 44
completeness of, 45
multiplication of, 45
order of, 45
Dedekind, Richard, 11, 43, 52, 188
denseness, 48

derivative, *see* differential calculus
dichotomy laws, 117
differential calculus, 50, 51
chain rule, 52
for complex numbers, 72–75
for constructive reals, 122
for quaternions, 91–93
product rule, 52
using hyperreals, 161
disjunction
in classical logic, 104
in constructive logic, 106
disjunctive argument, 105
distance, 20, 21, 29, 61
distributivity, 13
divergence, 92
dominated option, 183
Dominos, 176, 177, 184, 186, 187, 201, 202
dot product, 80, 81
double negation, *see also* pure existence proof
in classical logic, 105
in constructive logic, 107
double numbers, 69
dual numbers, 69
dyadic rationals, 194

embedding, 21
of the integers in any ordered field, 23
of the integers in the rationals, 10
of the natural numbers in any ordered field, 21
of the rational numbers in any ordered field, 47
of the rational numbers in the reals, 40

Index

of the reals in the hyperreals, 145
equivalence, 3, 5, 8, 12
   and congruence, 4, 6
   and equality, 3
   and similarity, 4
   equivalence classes, 5
   of Cantor real numbers, 39, 40
   of combinatorial games, 179–181
   of constructive real numbers, 109–111
   of hyperreal numbers, 138–141
   of surreal numbers, 189, 193
equivalence classes, 5, 8, 9, 36
   of sequences, 41
equivalence relation, 39
   in constructive logic, 108
Euler, Leonard, 101
existence statement
   in classical logic, 104
   in constructive logic, 107

Fermat numbers, 100, 101, 103
Fermat, Pierre, 100
field, 12, 18, 24, 35, 55
   Cauchy complete, 29
   cue-blob, 14, 19, 24
   of rational functions, 14, 19
   of rational numbers, 13
   order complete, 26, 29
   ordered, 27, 29
   skew, 77
formal languages, 126
   $\mathcal{L}$, 134–137
   examples, 130–132, 134
   formation rules, 129
   semantics of, 135
   sentence, 129
   structures for, 135
   symbols, 126
   syntax of, 127
   terms, 127
fractions, *see* rational numbers
fugitive sequences, 123, *see* weak counterexamples
functions
   complex, 70–72
   fundamental theorem of algebra, 66, 76
fuzzy games, 181

games, *see* combinatorial games
gap in a field, 49
Gauss, Carl Friederich, 11, 76
geometric algebra, 94
glb, *see* greatest lower bound
Goldbach conjecture, 116
gradient, 91
greatest integer, 26
greatest lower bound, 26

Hackenbush, 176, 177, 183, 186, 200, 202
   partizan, 200
Hamilton, William Rowan, 91, 94
Hankel, Hermann, 11
Hex, 202
hyperreal number system, 11, 27, 43, 125–170
   and formal languages, 126
   and the language $\mathcal{L}$, 134
   Archimedean property of, 152
   calculus using, 157, 159, 161
   completeness, 135
   continuity using, 159
   decimal expansion in, 146, 154–156

definition of, 138
embedded reals, 145
equality, 139
extension of real function to, 143
extension of real relations to, 142
hyperreal integers, 142
infinite numbers, 150
infinitesimal numbers, 150
limits in, 156, 157
need for an ultrafilter, 169
order, 139
real approximation, 157
satellite lines, 153
sequences, 156

imaginary part
   of a complex number, 58
impartial game, 187
implication
   in classical logic, 104
   in constructive logic, 106
incomparable
   elements, 162
   games, 179
induction
   mathematical, 21, 22
   transfinite, 164, 165
inductive, *see* recursive
inequalities, 19
   addition of, 20
   transitive law, 20
infinite descent, 174
infinite sequence, 28
infinitely large numbers, 27
   in the hyperreals, 142, 150–154
   in the surreals, 195
infinitesimals, 28, 125, 150, 153
   cloud, 152
   in the hyperreals, 125, 150–154
   in the surreals, 195
integers, 2, 23
integral domain property, 15
intermediate value theorem, 50, 53
   constructive view of, 120
   hyperreal proof, 159
isomorphism, 17
   field, 17
   order, 18

Kronecker, Leopold, 97, 101, 124

languages, *see* formal languages
Laplace's equation, 73
Laplacian, 91
least upper bound, 25, 26, 32, 33, 100
Leibniz, Gottfried, 125, 161
lightlike, 90
limits, 28–32
   by real approximation, 157
   classical definition of, 28
   complex, 61, 72
   hyperreal definition, 157
   using hyperreals, 156, 157
linear equations, 15
linear order, 18, 35, 55
   complex numbers and, 59
   for the constructive real numbers, 114–118
   for the surreal numbers, 190
logic, *see* classical logic, *see also* constructive logic
   proof-based, 106, 108
   truth-based, 103–106
Lorentz transformation, 87, 88, 91
Los' theorem, 147

Index

lub, *see* least upper bound

mathematical induction, 21–23
maximum value theorem, 50, 51
    constructive view of, 122
    hyperreal version, 160
metric completeness, *see* Cauchy completeness
Michelson-Morley experiment, 35
Minkowski separation, 89
modulus, 60
    of a complex number, 60
    of continuity, 120
    of differentiability, 122
Modus ponens, 105
Modus tollens, 105
monotonic, 29
    sequence, 53
multiplication
    of fractions, 10
    of pure quaternions, 81
    of reals, 12
    of surreal numbers, 190–192

nabla operator, 91
natural numbers, 23
negation
    in classical logic, 104
    in constructive logic, 106
negative of a game, 187
Newton's Method, 98
Nim, 176
norm
    Minkowski, 86
    of a complex quaternion, 86
    of a quaternion, 78
null
    cut, 44
    sequence, 37

numbers
    complex, 57–66, 76
    constructive, 97, 109–119, 124
    double, 69
    dual, 69
    hyperreal, 125–170
    quaternions, 77–94
    real, 35–46
    surreal, 171, 188–204

operation, 108
options, *see* combinatorial games
    of a game, 172
order, 18, 19
    completeness, 24–33
    isomorphism, 18
    of games, 178
    of the Cantor real numbers, 41
    partial, 162
    well, 163, 164
ordered field, 18, 23, 27, 29

partial order, 162
    incomparable elements, 162
partizan game, 187
Peano, Giuseppe, 11, 22, 53
Platonism, 103
Poincaré, Jules, 124
polar form
    of a complex number, 62
    of a complex quaternion, 87
    of a quaternion, 82
position
    of a game, 172
positive cut, 45
preferential ordering of games, 178
prefilter, 167
problem of definition, 9, 10
product rule, 52, 161

complex, 72
proof, 5
   as a basis for logic, 106
   by bisection, 31–33, 39, 51
   by contradiction, 16, 102, 104, 123
   by contrapositive, 25, 105
   by infinite descent, 174
   by mathematical induction, 21, 22, 102
   by playing a game, 185
   by transfer principle, 149
   by transfinite induction, 164, 165
   by trick, 39
   constructive, 111
   how to find, 16
   of existence, 102
   pure existence, 102
   what is one, 6
proof by bisection
   constructive criticism of, 122
property
   additive identity, 13
   additive inverse, 7, 13
   anti-symmetry, 162
   Archimedean, 27
   associativity, 12
   Cauchy complete, 29
   commutativity, 13
   conformality, 74
   congruence, 4, 6
   denseness, 48
   distributivity, 13
   multiplicative identity, 13
   multiplicative inverse, 13
   order completeness, 26
   reflexive, 3
   similarity, 4
   symmetry, 3
   transitivity, 3
   trichotomy law, 18
pure existence proof, 102
pure quaternion, 78, 80
   multiplication, 81

quadratic equations, 15
quadrupole field, 76
quantifier, *see also* existence statement, *see* universal statement
quaternion function, 92
   derivative, 93
quaternions, 11, 77–94
   addition of, 78
   calculus of, 92, 93
   Cartesian form of, 78
   conjugate of, 78
   differential calculus of, 91, 92
   multiplication of, 78, 79
   norm of, 78
   polar form, 82
   pure, 78, 80
   scalar part of, 78
   uniqueness of, 85
   unit, 78, 80
   vector part of, 78

rational function, 14
rational numbers, 2, 7, 10, 11, 13, 19, 24, 25, 28, 36, 53
   dyadic, 194
real numbers, 35–46, 98
   rational approximation to, 112
real part
   of a complex number, 58
recursive
   definition, 128
   proof, 128
reflection

Index

in space, 85
reflexivity, 3, 4
regular sequence, 109, 113
relativistic intervals
    lightlike, 90
    spacelike, 90
    timelike, 90
relativistic space-time, 89
Robinson, Abraham, 125, 170
rotation, 88
    axis, 83
    hyperbolic, 88
    in space, 82, 83

scalar part
    of a complex quaternion, 86
    of a quaternion, 78
sequence, 28, 29, 37
    bounded, 29
    Cauchy, 28, 29, 38
    defining hyperreals, 138
    infinite, 28
    monotonic, 29
    null, 37
    of hyperreals, 156
    positive, 41
    regular, 109
short, 201
simplicity theorem, 193, 194
skew field, 77, 85, 86
Snort, 177, 184, 186, 201, 202
space separation, 89
spacelike, 90
square roots
    of complex numbers, 65, 75
    of surreal numbers, 199
stopping value, 201
    left and right, 202
structures, 135

surreal number system, 11, 27, 171, 188–204
    birth order, 194
    day ω, 195
    definition, 188
    inverses in, 197–199
    linear order of, 190
    multiplication in, 190–192
    reciprocals in, 197
    simplicity theorem, 193, 194
    square roots in, 199
symmetry, 3, 4

Tarski, Alfred, 168
term, 127
the language $\mathcal{L}$, 134–137
the language $\mathcal{L}^*$, 147
theory of relativity, 89
time separation, 89
timelike, 90
transfer principle, 142, 148, 149
    proof by, 149
transfinite induction, 164, 165, 167
    Banach-Tarski paradox and, 168
transitivity, 4
    in logic, 105, 107
    of inequalities, 20
    property, 3
triangle inequality, 20, 37
trichotomy, 162, 163
    and combinatorial games, 179
    and inequalities, 19
    and the constructive real numbers, 114–117
    for surreal numbers, 190
    for the Cantor real numbers, 42
    in an ordered field, 18
truth
    as a basis for logic, 104

as a Platonic concept, 103
determined by convenience, 142, 143

ultrafilter, 140, 141, 166
    and the well-ordering principle, 166
    existence of, 167
    free, 140
    necessity of, 169
uniform continuity, 120
uniqueness
    of the complex numbers, 66
    of the quaternions, 85
unit quaternion, 78, 80
universal statement
    in classical logic, 104
    in constructive logic, 107
upper bound, 25, 26, 30, 33

vector part
    of a complex quaternion, 86
    of a quaternion, 78
vector space
    existence of a basis, 99

weak counterexamples, 101, 116, 121, 123
Weber, Heinrich Martin, 11
well-defined, 9, 10, 40, 41, 113
well-ordered set, 163, 164
well-ordering principle, 125, 165, 166
what is a proof, 6

zen buddhism, 6
zero
    as a game, 173
    definition of, 13

About the Author

Michael Henle was born in Washington, DC in 1944 and grew up in Arlinton, in the Commonwealth of Virigina. He was educated at Washington-Lee High School and then at Swarthmore College (1961–1965, BA in Mathematics). He did his graduate work at Yale receiving three degrees (MA 1976, MPhil 1969, Ph.D. 1970).

He has taught at Oberlin College since Fall 1970 as a Professor of Mathematics and Computer Science.

He has published a number of papers, including one in the *College Mathematics Journal* and the *Mathematics Magazine*. He has also published two books, *A Combinatorial Introduction to Topology* (San Francisco: W.H. Freeman and Co. 1978, reissued by Dover Publications 1994) and *Modern Geometries: The Analytic Approach* (Upper Saddle River: Prentice-Hall, Inc. 1996). The second edition of the latter (2001) has a slightly different title: *Modern Geometries: Non-Euclidean, Projective and Discrete*.

He is currently the editor of the *The College Mathematics Journal*.